Praise for Jay Ingram

"Lively ... witty ... Ingram's refreshing voice transforms mysteries into compelling reading. In the tradition of Stephen Jay Gould and Oliver Sacks, [Ingram] manages a difficult trick—making the minutiae of science seem alluring to the uninitiated." —*Maclean's*

"Ingram's eclecticism and his light, unmannered writing style are certainly a delight. But it is the freshness of his choices that I most appreciated ... in Ingram's hands, even the most unpromising material comes to life." —*New Scientist*

"Especially welcome, and extremely rare.... The quality and scope of Ingram's writing will lure many readers into scientific literature...." —*The Ottawa Citizen*

"Simplicity and wit ... everyone can relate to this refreshing look at science." —*The New York Times Book Review*

"Jay Ingram has taken on the role of alchemist: he turns the lead of scientific jargon into literary gold." —*The Calgary Herald*

"For those who like to ingest science in the form of intellectual hors d'oeuvres." —*The Washington Post*

"Ingram ... acts as a kind of cross between a clear-eyed journalist and a tongue-in-cheek comedian." —*The Globe and Mail*

"Ingram ... ably translates the nitty-gritty of science into engaging narrative, peppering his illuminating anecdotes with striking observations." —The *Gazette* (Montreal)

"Ingram knows very well how to turn technicalities into treats." —*The Toronto Star*

Also by Jay Ingram

The Science of Everyday Life

Talk, Talk, Talk

The Burning House

The Barmaid's Brain

For Children

Twins: An Amazing Investigation

Real Live Science

A Kid's Guide to the Brain

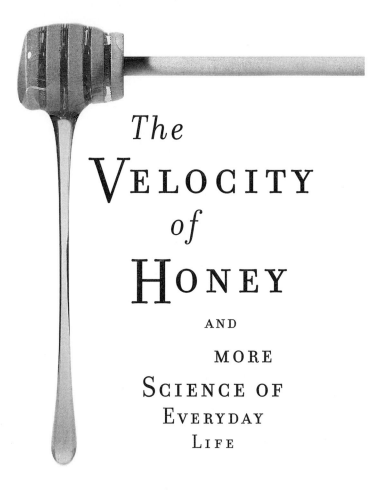

The
VELOCITY
of
HONEY

AND

MORE
SCIENCE OF
EVERYDAY
LIFE

JAY INGRAM

THUNDER'S MOUTH PRESS
NEW YORK

THE VELOCITY OF HONEY
and More Science of Everyday Life

Published by
Thunder's Mouth Press
An Imprint of Avalon Publishing Group Inc.
245 West 17th St., 11th Floor
New York, NY 10011

AVALON
publishing group incorporated

Originally published by Viking Canada in 2003
First Thunder's Mouth Press edition, February 2005

Library of Congress Cataloging-in-Publication Data is available.

ISBN 1-56025-654-0

9 8 7 6 5 4 3 2 1

Book design by Viking Canada
Printed in the United States of America
Distributed by Publishers Group West

To Cynthia, Rachel, Amelia and Max

Contents

Prologue:
Regis Was Wrong—There Is No Final Answer

This book is based entirely on the premise that "There's got to be more to it than that," "it" being any one of the thousands of moments that pass by unnoticed. Our lives are full of routine, much of which we are not even aware of, some that we have the good sense to ignore. But consequently we miss much—the little, apparently unremarkable events that, with a little digging, turn out to have charm and intrigue. That is the science of everyday life.

The word "science" inevitably raises expectations that answers or explanations will soon be delivered. This is a false expectation, based at least partly on how science is taught in school: there is no experiment performed in the lab, no page of a science text that fails to deliver a set of "facts." But science in the classroom bears little resemblance to real science in the real world, where there are very few, if any, final answers. If you're thinking of reading this book to find out those answers, I'm afraid you'll be disappointed. If instead you're hoping to gather some insight into the ongoing mysteries of science, then read on.

Science makes you pause, dig beneath the surface and think about things you'd ordinarily ignore. Is this a good thing? Of course it is. What kind of life is it to hurry like Alice's rabbit through the daily happenings without pausing and giving a little extra thought to any of them? Pausing to think has two obvious benefits: one, it guarantees at least a brief interruption in the ridiculous rush of life, which in turn is good for your mental and physical health, and two, sometimes you can't help but laugh at the irony of how little we know about what's right in front of our eyes. So think of this as a self-help book, a series of essays that reduce stress and restore a sense of being "in the moment." (If only it sells like a self-help book!)

I chose the topics based on the appeal of their science, which really meant their appeal *for me*. (For some unknown reason, there seems to be a lot of psychology and physics, with not much in between.) I don't pretend that this is anything more than a collection of experiences that we all occasionally share and recognize enough to say, "Yes, I've noticed that" (and at least one—echolocation—that most of us likely have never noticed). They are not, in most cases, the kind of research that makes it onto *Science* magazine's top ten advances of the year or wins Nobel prizes. But these pieces will make you think a little more about your life—if you let them.

The
VELOCITY
of
HONEY

AND

MORE

SCIENCE OF
EVERYDAY
LIFE

The Weird Physics
of the
Extremely
Ordinary

It's a truism that there's physics everywhere you look. After all, everything we're familiar with is made of atoms and molecules, and all their interactions are subject to forces that physicists have measured and described. Of course that may not include the mysterious events of parapsychology, but until such phenomena are actually shown to exist, it doesn't make much sense to speculate on whether or not they're physical or nonphysical.

But saying that everything is physics sounds more like the plaintive cry of physicists who aren't getting enough attention than an enticement to investigate further. On the other hand, if you said, "There's *weird* physics everywhere," people would want to hear more. If you then added, "There's weird physics in the most unexpected places," you'd have a captive audience. (Even better would be weird physics and sex, but such is the life of a science writer that this option is rarely available.)

So that's what I'm claiming: weird physics is right under your nose, and the example I'm using is crumpled paper. Take an ordinary sheet of 8 1/2-by-11 paper and scrunch it up into a ball. Really exert yourself to make that ball as small as possible. You may look at the ball as it rests in your palm with some satisfaction, feeling that you've accomplished something significant, but have you? No matter how hard you squeezed, at least 75 percent of that ball of paper is still air. Something in the crumpling process is resisting your best efforts.

Fortunately, some clever experiments at the University of Chicago have taken at least some of the mystery out of crumpling. The James Franck Institute at the university is peopled by scientists who take delight in working with simple materials and asking simple questions of them. Somehow, at least in the hands of these scientists, simple questions generate intriguing answers. For instance, to figure out what's going on when paper crumples, a group of the Chicago scientists (among them Sid Nagel and Tom Witten) devised a simple experiment.

They began by stuffing a 34-centimetre-diameter sheet of Mylar (the trade name for a kind of thin polyester film) into a tall cylinder, then placing a 200-gram weight on top of the Mylar. As you might expect, the weight squeezes the Mylar down—it crumples it. But maybe not the way you'd expect. This crumpling rivals a three-and-a-half-hour baseball game for tediousness. Rather than happening all at once, the weight continues to push the Mylar down for minutes, hours, days . . . even weeks. After three weeks the scientists got tired of the experiment, even though at that point there was still no sign that the crumpling process had stopped.

But that wasn't the most amazing aspect of this experiment. After the weight had been sitting on the Mylar for a little over eight minutes, the experimenters lifted it off for about another eight minutes (exactly five hundred seconds), allowing the Mylar to regain *most* of its former volume. Then they replaced the weight. What was truly bizarre was that the weight collapsed the Mylar down to where it would have been if the weight had never been removed. It was as if the Mylar held a memory of where it had been crumpled to. One further twist was added when the cylinder was vibrated back and forth seven and a half times a second: the crumpling accelerated noticeably, but still gave no signs of reaching an end.

What to make of all this? Scientists at the James Franck Institute have been at this for a while. They found out in the 1990s that any sheet of material, paper, plastic wrap or the old favourite, Mylar, seems to crumple in much the same way as long as it is large and thin. It really is better described as "buckling" than crumpling, because sharply angled ridges are created when pressure is applied, and the physicists have shown that the energy you've expended (or at least that part of it that ends up being stored in the sheet) is mostly contained—about 80 percent worth—in these ridges. They store energy like springs, and can give it back: any sheet that you squeeze as tightly as you can in your fist will start to unfold—with abrupt spasmodic movements—as soon as you open your hand.

Obviously the more crumpling, the more ridges, or creases, have to be created. But as the number of ridges increases, the force necessary to create even more rises steeply. To decrease the diameter of a crumpled ball of paper by two, you have to apply a force sixty-four times larger than was necessary to get it to its initial crumpled state.

This sounds paradoxical: the force necessary to squeeze the paper rises steadily, but in the lab the same weight continues to do the job for at least three weeks. What's happening? It must be that the energy that's stored in the ridges is gradually leaking away, but whether it's lost by rubbing against other folds in the paper or drained by the creation of permanent folds, nobody is sure.

It turns out the opposite process, that of unfolding a crumpled sheet, is just as interesting. Not so much from the energetic point of view but from the acoustic, because this is the science that addresses the annoying crackling that the people next to you at a concert make when they unwrap a candy at exactly the wrong time. While this is primarily an issue for abnormal psychology ("Why do that *now*?"), it's also physics. I don't know whether psychologists have investigated the phenomenon, but physicists have. And guess what? They were at the James Franck Institute.

Eric Kramer and Alexander Lobkovsky used the familiar standby in these experiments, Mylar. But in this case, they folded a sheet of the stuff thirty to forty times to create large numbers of permanent creases. They then made high-resolution digital recordings of the sounds emitted by the crumpled Mylar as it was unwrapped in front of a microphone in a music studio.

They heard the familiar crackling all right, but in this controlled situation they were able to show that the sounds were actually extremely brief clicks, each lasting about a hundredth of a second. It didn't seem to matter how thick the Mylar was: any sheet from a hundredth to a thousandth of an inch thick produced the same abbreviated bursts of sound when unwrapped. Although the duration of the sounds was astonishingly consistent, the energies were anything but. There was at least a millionfold range in energy from the smallest to the largest (or the faintest to the loudest).

Imagine a fresh new uncreased sheet of plastic. No matter how you bend it or roll it up, it snaps back to its original flat configuration. But as soon as you apply enough force to put creases in it, it can never return to its original form. Instead, the now crumpled piece of plastic has many stable configurations. When you roll it up into a ball and let it go, it relaxes a little, but still remains rolled up. Flatten it out and it may twitch a little, but it remains flat. Crumpled plastic can assume literally thousands of stable shapes.

Here's where the sounds come in. If you begin to unwrap the crumpled sheet of plastic, you are putting energy into it. The plastic stores this energy until enough has accumulated that the sheet can snap into a new configuration. The sudden movement of the plastic creates the sudden noise. Multiply a single event like that by the hundreds of readjustments a heavily creased candy wrapper makes as it is opened and you get the all too familiar snap, crackle and pop.

The important news in this was that unfolding the candy wrapping slowly actually does nothing to attenuate the sound. The same amount of energy is contained in the ridges, and whether they release that energy every once in a while or all at once, you're still going to generate the same amount of sound. So if someone has already made the choice to unwrap, doing it slowly or quickly simply means the sound is either prolonged or sudden. There's no way of lessening it.

There is a way of making more noise, though: by repeatedly bending the wrapping back and forth, it's possible to buckle and unbuckle the sheet in exactly the same place, creating a click each time. I hope no one would be perverse enough to do this in public.

For those who read this and wonder why these scientists don't get a life, I could include all the examples of crumpling that are poorly understood and will benefit from this work, such as the behaviour of car bumpers in collisions, the contortions red blood cells go through as they squeeze their way through the narrowest capillaries in our bodies, or even the folding of giant mountain ranges like the Himalayas. When adjusted for scale, all look a lot like crumpled paper (think of satellite photos of mountain ranges), and all are thin sheets buckling under stress, whether the sheet is the cell membrane of the red blood cell or the plates of the Earth's crust.

If you doubt that research like this affects people's lives, here's a story that will put those doubts to rest. A man named John Brunkhart, science news editor at America Online, read about the crumpling research and submitted the following memoir to the crumpling web page at the James Franck Institute.

"For whatever reason, for a week in the sixth or seventh grade, a number of my friends and I got into a competition to see who could fold a sheet of common notebook paper into the smallest volume. We tried many ways of folding the paper crisply, in many different patterns, and even crumpling. Ultimately, the winner was my friend Joe, who proved persistence, and not force, was the key.

"He began with a common crumpled sheet of paper, and then began pressing it with the thumb and forefinger of each hand, alternating hands, slowly compressing the sheet in the shape of a cube. After a brief time, it seemed the little cube of paper was small, but not getting any smaller. Joe, however, kept at it, continuing to apply force to the paper almost as a meditative exercise (no doubt to endure the more tedious classes). It soon became clear that the cube was slowly but surely getting smaller, even though Joe was applying no greater force at the end than at the beginning.

"By the end of a week of constant pressing, the cube of paper was truly tiny, perhaps a quarter of an inch to a side. We didn't believe it. Of course, we tried pressing our own sheets of paper together in the same manner with greater force, but we did not have Joe's patience to do it for as long, and never achieved the tiny, dense little cube that he did. For a time, I wondered how small Joe could have reduced his cube, if he kept at it in ensuing weeks and months. Even at ever-diminishing returns, it almost seemed as if he could keep compressing it forever."

Here you have it: crumpling research either sheds light on the silly (but memorable) activities we once pursued or it might have larger implications. As far as I'm concerned, while I'm confident that it will lead to important applications, for the moment I'm happy just to know a little more about a ball of crumpled paper.

The Velocity
of Honey

These next three chapters should be read over breakfast, and would even be enhanced by some activity. Not experimentation, just some straightforward observations with the bonus being that once you're finished you can eat—and drink—what you've watched. All you need is some toast, liquid honey, a spoon or one of those honey servers with the grooves on the end, and a cup of coffee. This is a highly sweetened version of the last physics course you ever took.

Start by putting a single slice of toast on a plate, then scoop up some honey, hold it just above the toast and watch closely as you slowly raise the spoon. The honey flows in a thin stream down onto the toast, but no sooner do you raise it even slightly than the stream begins to behave in a way that is somehow familiar, but somehow not. The honey doesn't just flow out in all directions across the toast, then the plate, the way water would. Honey is too viscous—too resistant to flow—so it begins to pile up at the spot where it's hitting the toast. As soon as that little cone-shaped pile forms, the stream of honey hits a roadblock. No longer able to fall directly to the toast, it instead strikes the top of the cone and then begins to fall down the slope. But that bends the stream and creates powerful forces that seek to bring the stream back into line.

There is usually one of two results. If you're using an implement that allows the stream of falling honey to be cylindrical, then the honey starts to coil on the toast as if it were a rope. If on the other hand you're using something that creates a ribbon of honey (pouring it from a wide-mouthed jar, for instance), then the ribbon, instead of coiling, flips back and forth as it lands. There are irregularities, especially if your hand wobbles, so the coils or flip-flops will wander around on the surface of the cone—the stream is exquisitely responsive to changes in its starting position.

The physics of all this becomes much more apparent if you move your hand up or down. Lower the spoon until it's just above the toast and you'll see that the coiling or flip-flopping stops. It's a question of gravity: if the honey hits the cone before it has built up an appreciable velocity, it is travelling slowly enough that, viscous as it is, it has time to flow away before piling up. However, raising the spoon only slightly will immediately trigger coiling again. (If you felt like it, you could calculate just what velocity the honey has to be travelling for it to cross that threshold.)

There continues to be a relationship between the height of the spoon and the pattern of coiling as you raise the spoon ever higher, but it's a little different: the higher the spoon, the faster the action when the honey hits. When scientists first experimented with this phenomenon in the late 1950s, they used transmission oil rather than honey (it's a lot easier to be sure you're getting much the same stuff from one batch to the next) and found that at a height of 9 centimetres, the stream laid down 120 coils a second, but if the stream fell from a much higher point, 18 centimetres, the rate was more than 300 coils per second. These coiling rates are pretty fantastic and would be impossible for the eye to follow, but with the honey falling from an extremely low altitude—just above the place where coiling disappears altogether—you can usually time the speed of coiling with the naked eye.

If the stream is cylindrical or tubular, the stream starts to coil when it's piling up on the toast because it is viscous enough to act as if it's a solid, like a rope. A rope coils as you let it down to the ground because that's the path of least resistance: you can't imagine a rope flipping back and forth because that would demand that it fold back on itself too abruptly. On the other hand, if the honey is streaming like a ribbon, flipping back and forth on landing requires less energy than coiling.

A completely different phenomenon kicks in when you raise the honey spoon so high that the stream of honey thins to the point where it actually separates from the spoon and falls in drops. Sidney Nagel and his colleagues at the University of Chicago have investigated exactly what happens when the stream thins and breaks, and they've found that what may look ordinary on the surface is in fact a most bizarre series of events. Unfortunately, these events are microscopic, so you will just have to take my word for it that they happen.

If we were watching water trickle onto toast, the dynamics would be very different: in the old familiar way, a drop would grow in size, then suddenly detach and fall to the toast. There would be no stringiness, no sticky strands. The fact that a drop forms at all is because water has surface tension, the tendency of all the water molecules at the surface to be pulled inward by their fellow molecules. This has the effect of creating a skin around a drop, and that skin allows the drop to grow in size until its weight exceeds the strength of the surface tension, at which point it separates and falls.

But the details of that process are curious. The drop doesn't just swell, then start to pull away, tapering the connection between it and, say, the tap, until it breaks away. The drop does grow, the connection does thin, but it continues to lengthen until the point where the thin strand of water still holding the drop in place is two or even three times as long as the drop itself. The drop is, at the last millisecond, almost spherical; the strand looks like nothing more than a needle inserted into the drop.

A liquid like honey that is much more viscous than water can sustain a much longer neck on a drop before breakaway. But there is something else as well. Photographs of these viscous drops just before they fall reveal an amazing thing: the strand still has that needle-like appearance where it meets the drop, but between the apparent point of the needle and the drop there is another, much thinner, strand connecting the two. It's as if a sixteen-lane highway narrowed suddenly to a back alley. The drop is still connected, but by a strand of liquid that is invisible to the naked eye. Then, as that fine strand lengthens and is just about to break, another, much narrower, appears—the back lane turned into a footpath. Everything is still connected (although the end is soon to come), with the drop attached to a strand that forms a neck to a slimmer strand, which forms a neck to a slimmer strand . . . and it may actually continue ad infinitum, or at least until the final strand is only a molecule or two thick. Each new strand is slimmer and forms faster than all the others. Even photos taken one ten-thousandth of a second before the drop falls show the series of necks intact.

When the scientists who did this work described it in an article in the journal *Science,* they called the process "a cascade of structure," practically poetic for an article in *Science.* They suspect that noise—in the sense of disturbing vibrations—may create this weird effect, but they also are pretty sure that no natural situation exists that would be without such disturbances. Certainly that would never be true in the kitchen, full of air currents, sounds, islands of heat and cold, and hands trembling with hunger, and that makes this entire trip of the honey from beginning to end, from pouring to eating, a very strange one indeed.

TUMBLING
TOAST

Experimenting with streams of honey usually leaves you with numerous pieces of sodden toast, and finding uses for them isn't easy. If they aren't too soaking, why not push them off the edge of the table? That is admittedly trickier if you're in a restaurant rather than your own kitchen, but it will illustrate yet another kitchen event that cries out for a scientific explanation: why does toast always fall butter side down?

I should point out that this is not simply a case of toast falling like a coin, sometimes one way, sometimes another, with hungry observers remembering only the unfortunate butter-side-down events. It is a real event. Also, it's not the butter itself: honey, marmalade, anchovy paste—they'll all work. Toast will fall spread side down almost all the time, and while some people have attempted to dismiss this as just another example of Murphy's law (what can go wrong will go wrong), analysis and experiment have shown that it is much more interesting than that.

It's important to set the parameters of the event first. The fall of the toast from the table is usually not a violent event; rather than being swept from the table (although a carelessly enthusiastic gesture with the arm might do that), buttered toast is usually tipped off the table edge, or slips from a plate as it's being set down. Toast falls; it rarely flies. This pretty well takes the phenomenon out of the realm of aerodynamics and into the more straightforward arena of gravity.

Let's assume the toast is sitting near the edge of the table—maybe even protruding slightly over the edge—when it's given a nudge. As more and more of the slice moves out over the edge of the table, its centre of gravity moves closer and closer to the edge. It's that centre of gravity, the place where physicists consider the mass of the toast to be concentrated, that is the key. Once that point extends out into the open space beyond the table edge, the toast is doomed. But it doesn't just fall flat. It starts to tip over—that is, it begins to *rotate*—and continues to do so until it hits the floor. It's that rotation that is the secret to the butter-side-down phenomenon.

Several groups of experimenters have tackled this problem. Some have used real toast. Others, frustrated by the inherent variability from one slice to the next, its crumbly nature and its tendency to become dry and brittle over time, have resorted instead to toast-like objects over which they can exert more control, such as a piece of similar-sized plywood. Bread samples have been obtained from, among others, the Alfred Nickles Bakery in Navarre, Ohio, and Michael Cain and Company in Oxford, England. Falling slices of toast have been videotaped, the numbers have been crunched and recrunched, and here is the gist of what's been found.

Robert Matthews, a science journalist who also holds a position at the University of Aston in England, published his analysis of falling toast in the *European Journal of Physics* in 1995. Matthews first pointed out that the butter (or whatever spread is used) is itself not a factor. The amount of butter—he estimated about 4 grams—is small compared with the typical 35 grams of toast and so cannot contribute significantly to the rotational mechanics of the falling bread. Also, the layer of butter isn't enough to disturb the flight of the toast once it has left the table, except in the unusual circumstance of the table being at least ten times higher than it typically is. Only in that case could the thin layer of butter disturb the flow of air around the toast significantly. In essence, then, the butter serves only as a marker for the top of the toast and doesn't actually play a role in which side ends facing up. Unbuttered toast behaves the same way (except that it could be brushed off and eaten if you were desperate).

Matthews focused on the tendency of the toast to rotate after it leaves the table. As I mentioned before, rotation is the key. Toast doesn't flutter to the floor like a leaf; it turns, and unless it turns less than 90 degrees, or more than 270 degrees, it's going to land butter side down. The question then becomes, how much rotation can be expected from a typical piece of toast? Matthews made some critical assumptions: that the toast was not moving horizontally when it left the table, and that it wouldn't bounce when it hit the floor. He then presented a series of equations that established how the toast would rotate depending on how far it hung over the edge of the table before finally tipping.

This relationship between overhang and rotation is critical, because the equations point out that if the toast can extend far enough out over the table edge before tipping it might just rotate fast enough to complete a turn before impact, thus landing butter side up. But is that degree of overhang possible?

Apparently not. Picture a uniform piece of toast about 10 centimetres long balanced on the edge of the table so that exactly half, 5 centimetres' worth, is hanging over the abyss. For that toast to rotate fast enough to make a complete 360 before hitting the floor it must not tip over until it's pushed out another 6 percent of that hanging 5 centimetres, or about 3 millimetres. That may not sound like much, but it's beyond the capability of most pieces of toast. Matthews found that typically toast can't be pushed out any more than about 0.6 millimetres before it falls. That distance simply isn't enough to impart rotation rapid enough that the toast will be saved.

These data may not apply to every single example of falling toast. For instance, Matthews built in the assumption that the toast would have no horizontal velocity, but of course if it has been struck by the sweep of an arm, it would have. The question is how much. He calculated that a piece of toast would have to be travelling something like 1.6 metres per second for it to sail across the room and land right side up. That's almost 6 kilometres an hour, a velocity that might just land it under (or even on!) the table next to you in a restaurant, a hefty swipe by any standards.

Matthews' article in the *European Journal of Physics* seemed to establish once and for all that it is no illusion: toast will almost always land butter side down. However, a more recent attempt to clarify the situation has clouded the picture somewhat. Michael Bacon and his colleagues at Thiel College in Greenville, Pennsylvania, performed their own set of experiments and went one step further than Matthews by taking into account the sliding of the toast off the edge of the table just after it tips and begins to fall. Their videotaped observations of a toast-sized piece of plywood showed that the free-falling "toast" was rotating much faster than it should have been, and they concluded that it was gaining that extra speed from that extremely brief interval of sliding.

How much difference this makes in the real-world version of tumbling toast is hard to say. Bacon et al.'s calculations showed that, if toast slippage is taken into account, a wide range of overhangs, from about 0.8 centimetres to 2.8 centimetres, would permit the toast to land butter side up. But to take advantage of this window of opportunity you would still have to get your toast to hang there suspended with its centre of gravity one or two centimetres out beyond the edge of the table before tipping, sliding and falling. As long as the toast overhang is less than 0.8 centimetres, the toast will fall butter side down, and, as the Bacon team wrote, "this overhang has to occur in almost all plausible accident scenarios (e.g., the toast is carelessly bumped off a tabletop) before tumbling takes place."

The Thiel scientists also wonder in their article in the *American Journal of Physics* whether bagels fall cream-cheese side down, and I can actually answer that. They do, but not always.

My daughter Amelia and her friend Lizzie Barrass completed a school science project in which they launched bagels off a table with varying degrees of force, and found that fast-flying bagels have a degree of aerodynamic stability: they will remain right side up if struck with the kind of force that might be equivalent to flicking one off the table with your hand.

They didn't determine whether this would have been true for Montreal bagels, bread, toast or any other vehicle for butter or cream cheese, or whether bagels, being disc-shaped with rounded edges, have some Frisbee-like aerodynamic properties that the others lack. Spinning discs do stay aloft nicely, and some of these bagels covered a lot of ground before landing.

Robert Matthews concludes his analysis of the matter by calculating that we would have to use tables three metres high to ensure that any toast that fell off would have the time to turn at least 270 degrees and so land butter side up. There have also been suggestions to reduce the size of the toast, which from the physics point of view would have the same effect as raising the height of the table, but toast squares about an inch across (2.5 centimetres) is what the theory demands, and those would be about as practical as a three-metre-high table. It has even been suggested that attaching the toast to a cat's back would work perfectly, because cats always land feet first.

Matthews suggests that quick reflexes might allow you to give a falling piece of toast a smack to ensure, like my daughter's bagels, that it sails across the room without rotating. But all these solutions fly in the face not only of practicality but of nature itself. Physicist William Press pointed out in 1980 that humans are the height they are for good reasons: we are more unstable than four-legged animals, and if we were any taller we would risk serious head injuries from a fall. So we are the height we are for good reasons, and our tables are the height they are for our comfort. As long as those factors hold true, toast *will* fall butter side down.

COFFEE
STAINS

The sad moment always comes when the congealed heaps of honey have been mopped up and the floor cleared of fallen toast— ah, but there's still coffee! It might be those late nights—and early mornings—in the lab, but whatever the reason, physicists are as entranced with the ordinary cup of coffee as they are with the other phenomena of the breakfast table.

If you can position a cup of hot, black coffee so that light strikes it at an angle, you should see a whitish sheen on the surface. (This works even better with a cup of clear tea.) There's something more to this sheen than first meets the eye. It makes a flagstone pattern on the surface of the coffee, with patches of this lighter colour separated from other patches by dark lines. The patches are usually a centimetre or so across.

These patches are what scientists call convection "cells," areas where warm fluid is rising and cold is sinking. Convection is what the weather is all about, not to mention ocean currents, and the same thing in miniature happens in your coffee. As the surface layer cools from contact with the air above it, it becomes denser and sinks, forcing warmer, less dense coffee up to the surface. But this doesn't happen in a haphazard or confusing way. Rather, the areas of upflow and downflow organize themselves into roughly similar sized columns, one beside the other. In the coffee cup the areas with the whitish sheen are rising columns of hot coffee, and it's the heat of that coffee that creates the sheen, although saying it in that straight-forward way misses the point: a drama is being played out at the surface of your coffee.

The sheen is actually a thin layer of tiny water droplets, droplets that have condensed just above the surface of the coffee and are hovering there, less than a millimetre above the surface. It's whitish because so much light reflects from the surfaces of the droplets. The droplets form because as the water evaporates from the hot surface of the liquid, it cools suddenly, condenses and coalesces. The drops that form do not fall back onto the surface of the coffee because they are buoyed up by the trillions of water molecules still rising up underneath them. Held there, suspended above the surface, they are clouds on a scale so minute that only careful lighting reveals them. It would be an incredible experience to be there in the tiny space under the droplets but above the liquid coffee. It would be hellish hot for one thing, but you'd also be buffeted by stuff evaporating from the surface, and concerned all the while about slipping into the downstream convection (the black lines separating the clouds) and vanishing into the blackness of the coffee below. Even from our mundane perspective (simply looking down on the cup) it should have been apparent from the start that the drops were hovering—you would have noticed that a breath scatters them instantly, like clouds before the wind, but they form again just as quickly.

The only place where you can see right down to the coffee surface is along the black lines, as if you are seeing the surface of Venus through a sudden break in its impenetrable clouds. The cool coffee sinks in those black lines, completing the convection cell.

Just as an aside, the same thing happens, believe it or not, in a jar of mixed nuts. Convection plays an important role in what has been called the Brazil Nut Effect, the phenomenon where the larger nuts in a collection end up near the top of the jar. There is more than one reason for this: one of the more prosaic is that smaller nuts or pieces of nuts can fall through narrow openings in the mix, whereas the larger nuts can't. But a far more interesting fact is that when the jar is shaken, a kind of convection flow is set up, whereby the nuts rise through the centre of the jar, move out to the sides when they've reached the top, then sink down along the sides to the bottom. The problem for the big nuts arises when they have reached the side of the jar and are being tugged down: the downward channel is too narrow for them to fit, so they stay on top.

Back to your cup of coffee: less beauteous than evanescent clouds or churning convection cells, but certainly more common, is the dark ring left behind when coffee spills. Even the ring presented a puzzle for physicists to solve. Funnily enough, when the puzzle was solved, the processes involved turned out to be the same as seen in coffee while it was still in the cup: the flagstone pattern and the clouds, the movement of fluid from one place to another and evaporation.

The puzzle is this: why, when a drop, or half a cup, of coffee spills and then dries, does it form a ring, with almost all of the dark coffee stuff in the ring and the centre almost empty? Why shouldn't it dry and leave a uniform beige stain on the counter?

Here are some clues: you can show that it doesn't have anything to do with gravity by throwing your cup of coffee onto the ceiling, then watching as it dries. Each individual drop will still form a dark ring at its perimeter. On the other hand, it must have something to do with evaporation, the process by which the water molecules move into the air, leaving the solids behind. A couple of early experiments by Sid Nagel and his colleagues at the James Franck Institute at the University of Chicago (again!) tested this by interfering with the normal evaporative process. In one, a drop was placed under a tiny glass lid that had only a minute hole over the very centre of the drop. You would expect under these conditions that the only evaporation possible would be from the centre of the drop, not from the edge. In this special circumstance drops did *not* leave a ring behind. So evaporation from the edge of the spill must have something to do with the formation of the ring.

In a second experiment the scientists placed drops on Teflon, to which, as you know, nothing sticks. Drops left on Teflon didn't leave a ring either. In this case you'd have to suppose that the smoothness of Teflon would be the key, suggesting that a second factor in ring formation is the surface on which the drop is sitting. Add to these the fact that if you use a microscope to watch the behaviour of tiny particles in the drop as it is drying, you'll see that the particles are streaming headlong out to the edge of the drop. Sid Nagel described it as being like watching rush hour in New York. Evaporation, the surface, the streaming—those are the things you need to know to be able to account for the ring.

Imagine looking at a drop—highly magnified—from the side. Each liquid will form a drop of slightly different shape, the determining feature being the surface tension of the liquid. A drop of a liquid like water behaves as if it's a liquid in a bag, as if the surface of the drop is acting as a skin to hold the rest of the water in. When you first learn about surface tension in school, it's attributed to attraction between the molecules of the liquid. In the middle of the drop a water molecule is pulled equally in all directions by its neighbours. But the surface, its upper side, is in contact with air, not water, so it's not being pulled in that direction. If its temperature is high enough, it might want to fly off into the air, but it's not being pulled in that direction; it's being pulled down by its fellow water molecules. When your tap is leaking, water doesn't fall as a continuous series of micro-drops, but rather forms sizable balls of water held together by surface tension until gravity takes over.

A drop of coffee, which is largely water, forms on the countertop under the influence of that surface tension. It doesn't flow out in all directions to form a film no more than a molecule thick—it balls itself up and sits there. Seen from the side it might have the profile of a one-person tent: flatter than an igloo but not much. The point where the edge of the drop meets the surface it's sitting on is critical: the angle the drop makes right there is not negotiable—it's a fundamental property of water.

So what happens when the drop is exposed to air and begins to evaporate? At the very edges, water molecules fly off into the air, thinning the edge. Ah, but the edge can't do that, because that changes the angle the drop makes with the surface, something surface tension will not allow. One option for the drop would be to retract its edges, pull them in to restore the original shape of the drop. But on normal surfaces, that can't happen, because the edge of the droplet is pinned to the surface by tiny irregularities. By this I don't mean that if you washed the kitchen counter every once in a while, the coffee spills wouldn't have rings around them. These are subtle, tiny, microscopic irregularities that exist in all materials except very special ones. On any normal surface, the edge of the drop can't retreat over the lumps and bumps, so the drop is trapped. There is then only one option: water must flow out from the centre of the drop to the edge to replenish that which has evaporated.

Of course, it's not just the water that vacates the centre of the drop for the edges. With it goes all the dissolved and particulate matter that exists in a cup of coffee. It is carried along, then finally dumped at the edge of the drop when all the water has evaporated.

The Teflon experiment worked because the surface is virtually free of irregularities, so the drop can contract as it evaporates, maintaining its preferred shape to the bitter end. The lid experiment worked because the water could not evaporate from the edges of the drop, only from the centre, so there was no need for the liquid to migrate out to the edges, no transport of particulate matter from the centre and therefore no ring. In that case the particulate matter was simply left where it was, forming a smudge.

Sip your coffee, gulp it, even spill it, but above all, take a second or two to check it out. After all, a glance at an apple stimulated great thoughts in Isaac Newton's head. It's true there aren't very many Newtons, but a few moments at the breakfast table can serve as a reminder that yes, our lives are under the influence of forces beyond our control: forces like surface tension, viscosity, evaporation and gravity.

Life's
Illusions

I love illusions of all kinds. The most common are visual illusions, the ones that used to be called "optical" illusions. The name was changed because most of them have nothing to do with optics, the interaction of sight with light. Instead they result from mistakes that your brain sometimes makes when trying to make sense of a complicated scene. It plays the odds, making the assumption that things in the world out there are behaving as they should, and sometimes gets fooled when they're not.

A good example is the moon illusion, the huge fat moon that hangs over the horizon shortly after rising (or just before setting). This is not an optical phenomenon caused by the light being bent or stretched by the atmosphere. You can prove it by comparing the size of the horizon moon with the overhead moon hours later, and you'll find that they match exactly. (One good way of doing this is to hold some kind of pill—an Aspirin is a good one—at arm's length. You'll see that it covers the moon just about perfectly no matter where it is in the sky.) There are several explanations for the moon illusion, the most popular being one that somehow your brain judges the distance to the horizon as being farther than the distance to the "roof" of the sky overhead. Then, confronted by the fact that the size of the image the moon makes on the retina of the eye is the same in both situations, your brain concludes that the moon must actually be bigger when it's at the horizon, and makes the appropriate correction to its apparent size.

I admit that there's something about this explanation that makes me uneasy. It seems a little convoluted, and there is good evidence that other factors, like the upward angle of your eyes or the tilt of your head and even the presence of mist or light cloud, play an important role as well.

There are illusions of sound and touch too, but because we are visual creatures—and have been for tens of millions of years—visual illusions predominate. But there's one problem with illusions: to be understood, most of them need to be carefully set up by an illustration in a book or computer animation. They are not something that you usually experience in your daily routine.

However, there are two illusions that I'm particularly fond of that are common, easy to experience and do what all good illusions should: give you a peek into your own brain's sometimes strange machinations.

The first is what I call the microwave illusion, although it also works just as well if you have a watch with a sweep second hand. (The microwave timer must be on and counting off the seconds.) If you are not looking at the microwave, then turn to it to see what time it is, your first impression will be that the counter seems to be frozen: it is as if the first tick is delayed. Then, once the ticking appears to resume, it carries on at the normal rate. Turn away again, then look back, and the same thing happens. The first change of the numerals (or in the case of the watch, the first movement of the second hand) takes much longer than the subsequent moves. If you try this repeatedly, the effect begins to wear off, perhaps because you are beginning to anticipate it or because of some as yet unidentified factor in the underlying psychology. But the beauty of this illusion is that you can abandon it at any time confident that after a break you will inevitably experience it again and again as you glance at your watch or the microwave.

Researchers in the U.K. first examined this weird phenomenon in 2001 and decided that the human brain was responsible for the time delay. They concluded that the brain covers up for the movement of the eyes as they shift from wherever they were originally focused to the clock. Rather than having you perceive that moment as a blur, the brain back-times the first image of the clock to cover the time the eyes were moving. And, the farther the eyes had to move, and the longer they take, the longer the apparent delay before the first tick of the clock. The U.K. scientists showed this by having volunteers participate in a simple lab test. They were to fix their eyes on a cross on one side of a computer screen, then switch their gaze to a zero on the other side. The rapid movement of the eye (called a "saccade") from one to the other took about a tenth of a second. At the moment the eyes started to move, the computer started to count off seconds, changing the zero to a one, then a two and so on. The fact that the timer started simultaneously with the initiation of the eye movement ensured that the volunteers would not see the zero for a full second (their eyes had to get there first) but would see each of the subsequent numbers for that long.

This experimental set-up had all the prerequisites of the illusion but with one advantage. The researchers could adjust the duration of the zero on the screen. Assuming that the volunteers thought it was lasting longer than the numbers that followed (they were experiencing the illusion), the experimenters could shorten the zero's life on the screen until it seemed to be lasting about as long as the other numbers. The amount it had been shortened would then be a measure of the "size" of the illusion.

The experiment was run with the cross and the zero two different distances apart. The larger distance of course would require a bigger eye movement that would take a little longer. It turned out that the bigger the eye movement, the more exaggerated the illusion: when participants needed to shift their eyes only 22 degrees (about half a 45-degree angle) across the screen, the actual duration of the zero had to be shortened to 880 thousandths of a second to make it seem as if it lasted a full second. In other words, the illusion had expanded the zero's apparent time on the screen by 120 thousandths, or a little more than a tenth, of a second. But if the eyes shifted across a much greater distance of 55 degrees, the zero needed to be visible for only 811 thousandths of a second, the illusion now lasting nearly two-tenths of a second. The difference between the two times—69 milliseconds—was almost exactly equal to the extra time needed for the eyes to reach the target: 67 milliseconds. The longer the tracking movements of the eyes take, the longer the illusion, because the illusion must cover up the blur. If that weren't evidence enough, a slight alteration of the experiment underlined the importance of the eye movements: if instead of the eyes moving to the target, the target moved to the eyes, the illusion disappeared.

In this instance the eye movements were conscious and directed toward a singular target, but they are not always like that. No matter what you're looking at at any moment, your eyes are constantly shifting from one feature of the object to another, and you are unaware of most of these movements. If you're staring at a face, for instance, your eyes are skittering from forehead to lips to nose to eyes to lips to nose, back and forth, pausing only for milliseconds at each stop. Such movements, even though you are unaware of them, are essential to your awareness of that face. Experiments decades ago with tiny contact lenses with images on them proved that. The trick with an image on a contact lens is that no matter how many saccades the eyes make, the image is stabilized over the lens. Because the eye is therefore forced to look at the same part of it, scanning it is impossible, with the result that slowly but surely, it disappears! This was one of the most spectacular pieces of evidence showing that saccades are a necessary part of visual perception.

As curious as saccades are, the conclusion that the brain is covering up for them by stopping time is even more curious. After all, if saccades are a standard part of the visual process, if they're happening all the time your eyes are open, then your brain is likely busy creating all kinds of stoppages in time to disguise them. It may not happen all the time—in the above experiment there was no illusion if the target shifted while the eyes were moving toward it. But if you're in familiar surroundings, the likelihood of objects not being where you expect them to be is probably slim, and therefore the illusion should be operating most of the time. Of course the reason you're not aware of the illusion is that you're rarely moving your eyes to a timepiece that would reveal it. When you do, it becomes obvious.

The idea that this illusion was all about eye movements made perfect sense on the basis of this experiment, but a second British research team, unrelated to the first, realized that there is a second version of this illusion that has nothing to do with vision. It happens when you are waiting for the phone to be answered. These days you usually don't have to wait too long before voice mail kicks in, but if the person you're calling is technologically challenged, you're stuck there, listening to ring after ring. If during that time you are distracted, either by holding the receiver away from your ear and talking to someone or by simply paying attention to something else, when you return to listening, the silent space before the next ring will seem much longer than any of those that follow. It's a close analogy to the microwave illusion, but it involves sound, not sight, and therefore can't have anything to do with saccades.

In this case the researchers used a similar experiment to demonstrate that they were working with a similar phenomenon: they distracted participants by having them listen to a series of tones in one ear, then switched them to the other ear to judge the length of silent spaces between a different set of tones. Most judged the first space to be longer than all the rest, although all were exactly one second long. Just as in the visual version, the experimenters were able to shrink the length of the first space until it seemed equal to the others: that length was about 825 milliseconds, again meaning that the space had to be shortened by more than a tenth of a second to make it appear to be consistent with the spaces that followed.

How is it possible to reconcile this result with the microwave illusion? Obviously eye saccades have nothing to do with the telephone illusion. The team who devised the auditory experiments has come up with a more general explanation. They explain the illusion in terms of our internal biological clock. They point out that experiments as much as forty years ago suggested that the ticking of that clock is variable. If nothing remarkable comes to the attention of the brain, the clock stays steady. But as soon as something novel occurs, the nervous system is aroused, and the ticking gets faster. This accelerated pace generally doesn't last very long—novelty is short-lived for our brains, and soon the clock returns to its original steady pace. But in that brief interval while the clock was speeding, the passage of time would have appeared to slow down. Obviously a biological clock doesn't literally tick (it depends on the rate of chemical reactions in our brains), but for the sake of argument let's imagine it does. If the clock doubled its speed, then a tick that formerly took a second would now take only half a second. That would mean that an event that used to last one tick (one second) would now take up two ticks (still one second) and would appear to have slowed down dramatically.

It sounds like a reasonable explanation, although there is still something about the microwave illusion that makes it seem like a visual trick. In that experiment the length of the illusion corresponded almost perfectly to the length of time the eyes took to move from one target to the next. Would that necessarily be true if it were simply a case of a temporary speeding of the biological clock?

The researchers who described the microwave illusion don't think so. They've come back with a new experiment that extends this illusion to yet another sense: touch. This experiment convinces them that attention is *not* the explanation for what's going on.

The touch experiment required the participants to reach to a target and put their hand on it. The target was vibrating when the person's hand made contact, but its pattern of vibration flipped back and forth between 120 times a second and 60 times a second. Each of those different vibrations lasted exactly one second, except the first, which was adjustable, just like the zero on the screen of the visual version of the experiment. Subjects were allowed to vary it until it seemed to last the same length of time as the other vibrations. As in that first experiment, the participants judged that first vibration to be equal in length to subsequent ones only when it was nearly a full tenth of a second shorter. But here's the important part: the hand had to have moved to create the effect. If it were already resting on the target when the vibration started, there was no illusion: in this situation people in the experiment adjusted the timing to something close to a full second. But as long as they had had to reach, they seemed to feel that their initial touch lasted much longer than it actually had.

So what sort of explanation can account for this, the third version of what the scientists are calling "chronostasis"—time standing still? The experimenters responsible for both the visual and touch versions are suggesting that what is happening here involves a little brain preparation. There's evidence from studies of monkeys that some brain cells begin to prepare for a saccade—a shift in eye direction—before it actually happens. These are cells that will be paying attention to the target once the eyes arrive at it. In other words, there are cells that seem to know where the eyes are heading and that get ready to focus their attention on that spot. Those are monkey brains, and the task is visual, but the researchers argue that the preparatory activity of those cells may be a common thread. Although they have no direct evidence, they suggest that reaching might well involve brain cells that similarly prepare themselves to respond to the touch of an intended target, and in doing so prepare the ground for—and are likely involved in—the illusion, but exactly how isn't yet known.

So far three of the five senses have been shown to exhibit this illusion of time, and I'd guess it's unlikely that taste or smell will be added to the list: they are a bit vague, a little too hard to measure or time, and timing is the key.

The experiments that have been done so far have required the presence of an accurate timepiece to reveal the illusion. But we have to assume that such illusions are happening constantly as we go through the day. Without a clock to reveal them, they are no longer illusions, just the seamless web of events that our brains like to create for us. The day doesn't feel as if it is filled with papered-over blurs of the eye, unduly long pauses before telephone rings or imaginary touches—*as far as we can tell*. But they must be there; we simply don't have any way of revealing them. And you thought you knew what was going on around you.

THE MYSTERIOUS ART
—AND SCIENCE—
OF BABY-HOLDING

W e are not symmetrical. Almost everyone is either right- or left-handed, true ambidexterity being very rare; babies turn their heads more to the right than the left, beginning in the womb; and people even preferentially turn their heads to the right, rather than the left, to kiss. But the most puzzling example is baby-holding. Most women prefer to cradle an infant on the left side of their bodies. Men share that preference, albeit much less strongly, but nearly half a century of research has yet to pin down the cause.

There is no shortage of suggestions for the behaviour, ranging from the claim (from the early nineteenth century) that a woman's left breast tends to be more sensitive than her right, prompting her to hold her baby's head on that side, to the idea that the emotional right hemispheres of both baby's and mother's brains are playing the key role. But theorizing is useless without data. Fortunately in this case there is plenty of that.

One of the largest bodies of data is, at the same time, one of the most mysterious. While you can find examples of mothers holding babies practically anywhere, one of the best sources is works of art from the past. At least three researchers have scoured catalogues of art and visited galleries, seeking to identify the baby-holding practices of the past, and their results are curious to say the least. Two of these studies focused primarily on Old World art, both paintings and sculpture, ranging from ancient Egypt and Greece to the twentieth century. Both found that women predominantly held their babies on the left side (many of these being the Madonna with the baby Jesus), but there were some puzzling changes over time. In a study published in 1975, Stanley Finger of Washington University found that the earliest works he examined, those of Giotto and Fra Angelico in the fourteenth and early fifteenth centuries, almost exclusively depicted left-side holding, but that this preference changed dramatically over the next half-century. So, for example, while 100 percent of twenty-six paintings by Fra Angelico (1387–1455) were left-sided, that percentage dropped dramatically to 38 percent of forty-two by Botticelli (1444–1510). Then for some unknown reason the left-side tendency was re-established by the beginning of the eighteenth century and has continued to the present.

A second study of both paintings and sculptures by Otto-Joachim Grusser at the Free University of Berlin came to similar conclusions. Left-holding was predominant in thirteenth- and fourteenth-century paintings, dropped to about 50 percent through the fifteenth and sixteenth centuries, then slowly rose again to hold at about 60 percent up to the twentieth century. Sculptures too were high in left-holding in the beginning, dropped to a low of about 50 percent at the end of the fifteenth century, then recovered by the beginning of the twentieth.

As you will see, these changes over time are very hard to explain with the current theories of why most women (and to a lesser extent men) hold babies on their left side. But inconsistency is not just seen in Western art. A third study, of ancient Central and South American art by Chilean neurologist Gonzalo Alvarez, showed that there were distinct regional differences. In the period between 300 B.C. and 600 A.D., cultures in Mexico and Central America were strongly predisposed to left-side holding—up to 83 percent—while at the same time in the Andes (now Colombia, Peru and Ecuador) that tendency was much weaker, barely over 50 percent. This is of course long before there was any contact with Europeans, and interestingly, also at a time of cultural upheaval in Mesoamerica.

These three studies together include about 2600 works of art, ranging over more than two millennia. There are some gaps: no works from China, Africa or India have yet been studied, and of course paintings and sculptures are not the equivalent of candid photographs. Because art is interpretive and subject to whim, fashion and even the will of the person commissioning the art, these works cannot be assumed to be a perfectly reliable record of how people in the past have carried their infants. Nonetheless, it is a large body of data, and when the theories of why women and men carry babies the way they do are taken into account, these works of art assume a much greater significance.

The first theory that really caught the attention of the public and researchers was put forward by psychologist Dr. Lee Salk (brother of polio-vaccine inventor Jonas Salk) in the late 1960s. He was inspired to think about the puzzle when watching a rhesus monkey holding her newborn infant (on the left side) in New York's Central Park Zoo. He wondered if the heart, or more specifically the sound of the heartbeat, was the secret—did women hold their newborns on the left side so that the infant could better hear the calming (and familiar) rhythm of her heart? Salk did his own informal observations of mothers with newborns, and found that of 255 right-handed mothers, 83 percent held their baby on the left. The surprising thing was that left-handed mothers were nearly the same: 78 percent. When Salk quizzed the mothers on their reasons for holding their babies the way they did, the left-handers replied that because they were left-handed they could hold their baby better that way; the right-handers claimed that doing it that way freed their right hand to do other things. Both were perfectly reasonable answers, but Salk suspected both were rationalizations for a behaviour that the mothers had likely performed unconsciously.

To probe the mystery deeper, Salk's colleague Hyman Weiland went to the supermarket. He positioned himself so that he could watch people leave the supermarket through an automatic door (eliminating the necessity for the shoppers to free one hand or the other for door-opening) and he studied only those carrying a single bag about the size of a baby. He found absolutely no preference for right or left hand among more than four hundred shoppers: the split was exactly fifty-fifty. This and other studies convinced Weiland and Salk that the left-side tendency had something to do with carrying a baby, not simply a package.

Salk had already done his own personal survey of art and had seen trends similar to those I mentioned earlier: left-side holding prevailed throughout the more than four hundred paintings and sculptures he examined. But it was a chance observation at a paediatric clinic in New York that convinced him of the heartbeat theory. He noticed that mothers who had given birth prematurely, and who as a result had been separated from their newborns for at least the first twenty-four hours after birth, were much less likely to hold their babies on the left. In fact a simple experiment—giving a baby to its mother by handing it to her exactly in the middle of her body and seeing on which side she placed it—confirmed that mothers who had had no postpartum separation from their babies were much more likely to hold their babies on the left compared with mothers who had been separated. Salk established that the effect had nothing to do with the prematurity of the baby, only with the fact that mother and baby had been separated.

Salk concluded that the sound of the mother's heartbeat must be the factor that prompted left-side holding. After all, it is a sound that babies hear in the womb, and would be one of the few steady and familiar stimuli present after birth. Although the heart lies more or less on the midline of the body, it torques slightly to the left, enough, in Salk's mind, to prompt mothers to hold their babies on the left side of their bodies. He then performed an experiment to support this idea: one group of babies in a hospital nursery was exposed to the recorded sound of a human heartbeat, the other had no sound piped into the nursery at all. (The original plan had been to expose the second group to the sound of a fast-beating heart or some other noise, but the babies became instantly distressed upon hearing those sounds, so that plan was abandoned.) After three days the babies who heard the heartbeat had gained more weight and cried less than the control group, convincing Salk not only that the heartbeat sound was an essential factor in the early mother–child relationship but that holding the child on the left side would make that heartbeat easier to hear. Apparently the mothers who had been separated from their babies for the first twenty-four hours had missed some sort of crucial milestone for establishing that behaviour in the first place.

As compelling as Salk's theory and the supporting experiments might seem, they were far from convincing to everyone. Some argued that the heartbeat can be heard just about as well by an infant being held on the right side; others felt that not all the alternatives had been exhausted. The idea that the left breast tends to be more sensitive didn't receive much support, but increasingly, psychologists entered the picture wondering if some sort of communication between mother and child was being facilitated by having the child's head on the mother's left side.

The psychology is based on our lopsidedness. Our brain is split pretty much right down the middle, but the two major hemispheres, left and right, process information differently. It's not that they are completely separate—don't be misled by self-improvement ploys that beguile you with claims that you can switch on your right hemisphere and become a poet or a tennis star overnight. It's not as if the two hemispheres are unaware of each other: they communicate instantaneously via a thick cable of neurons joining the two. However, it *is* true that each half has its specialties, a situation that has been revealed by studies of people who have had that inter-hemispheric cable cut to relieve the intensity of epileptic seizures. In the case of mother and child, the hemispheric specialty of interest is the processing of emotion-laden signals. When it comes to both facial expressions and tone of voice, there's good evidence that the right hemisphere is a little quicker, a little more adept at picking up the underlying emotion.

Here's an example. When psychologists briefly show people what are called "chimaeric" faces (one side with a neutral expression, the other portraying an emotion like anger or sadness), the majority claim the face is exhibiting the emotion being portrayed on the subject's left (the face's right). So a face that is neutral on its left side but angry on its right is judged to be an angry face. This bias toward identifying the emotion displayed on the right side of the face can be traced to the right hemisphere of the brain. To a large extent the wiring from each eye to the brain crosses over, so that images visible to the left side of each eye are primarily transmitted to the right hemisphere and vice versa. This is a momentary effect—obviously if you stare at a face long enough both hemispheres of the brain will be fully aware of it. But for a moment, the left side of whatever you see is conveyed to the right hemisphere and vice versa.

The same is true of hearing. What the left ear hears is sent primarily to the right hemisphere of the brain. "Dichotic listening" tests are the auditory equivalent of chimaeric faces, in that they provide evidence as to which hemisphere dominates when it comes to detecting emotion. And again, the right hemisphere is superior: if you listen through headphones to two sentences spoken simultaneously but with different emotional tones, the one heard by your left ear overrides the other. In other words, an angry tone in your left ear paired with a happy tone in the right will persuade you that the overall tone is angry, because your right hemisphere—to which the sound from your left ear has been routed—is more adept at detecting emotional tone. This effect is evident when listening to speech: the nuts-and-bolts of speech, the grammar, sentence structure and vocabulary, are decoded in the left hemisphere, but the flow and tone are processed in the right.

How might this cross-wiring play a role in baby-holding? Imagine you're holding a baby in your arms with the baby's head to your left, the position that the majority of women prefer. The baby's face is in your left visual and auditory field, and any facial expression or sound that the baby makes will be sent preferentially to your right hemisphere, the hemisphere that you would choose in that situation because it is so good at emotional detection. Right-side holding would, on the other hand, send emotional signals to the left hemisphere, a pretty good hemisphere as brains go, but not as good as the right when it comes to discerning emotions.

It's not just the perception of emotion but also its expression that is different in the two hemispheres. Each side of the brain controls the movements of the opposite side of the body, including the muscles of the face. The powerfully emotional right hemisphere again dominates, meaning that emotions are expressed more vividly on the left side of the face. (Note the irony: when you look at a face, you are most sensitive to the right half of that face, but the true emotion is expressed most strongly on the left side.) A baby held on the mother's left side would not only expose the left side of its face to the mother but would see the expressive side of the mother's face as well. It's a bit tricky to quiz infants on what's best for them, but it is possible to test mothers. So here goes.

John Manning of Liverpool University was one of the first investigators to subject these hemispheric ideas to experiment. He gathered together a group of four hundred girls between the ages of six and sixteen and tested their baby-holding preferences. Why girls of these ages who had never had babies? It had already been shown that girls show a left-side cradling tendency from the age of *four* onwards (a fascinating observation that demonstrates at the very least that this left-side holding is not triggered by having given birth). Manning divided the girls into groups, then tested whether having an eye patch over the left eye, right eye or both had any effect on how they held not a real baby but a realistic doll. The results supported the idea that brain hemispheres might have something to do with this. The group who had both eyes available favoured left-side holding by a wide margin: 80 percent. That percentage didn't change much in the group whose right eyes were blocked by the patch, but blocking the left eye had a dramatic effect: the tendency to hold the doll on the left fell to 61 percent, an even lower figure than the 66 percent of the totally blindfolded group. The most reasonable explanation of the results is that blocking the left eye interrupts visual communication with the doll and forces many of those girls to shift the doll to the unblocked right-hand side. However, even with that impediment, more than half still favoured the left side.

Manning followed up this experiment with one using mothers and their real babies. Each was asked to walk over to a crib where the baby was lying, pick the baby up and return to her chair and sit down. Again, eye patches were used (although they had to eliminate the totally blindfolded group for fairly obvious reasons), and this time the results were even more striking: with both eyes open the mothers held their babies on the left just over 60 percent of the time, but when the left eye was blocked, they switched to 60 percent right-holding. An oddity of this particular experiment was that the control group, with both eyes open, did not show the typical 75 to 80 percent preference for left-side holding. Manning suspected that was because the babies in this experiment ranged in age up to eleven months, and previous studies had shown that the maternal left-side tendency fades with time.

If you were to take only these experiments into account you could easily be convinced that vision and emotion together play the most important role in baby-holding, but subsequent experiments aimed at firming up those results have failed to demonstrate that link. One in particular, which closely duplicated the design of the Manning experiment, found that females weren't significantly affected by having their left eye blocked, but males were: they readily switched to right-holding when their left eye was blocked, suggesting, if anything, that males were more dependent on the sight of the baby (or in this case, doll) than females. But it's well known that males don't exhibit the left-side holding tendency nearly as strongly as females. If vision is so important for them, maybe it's less important for females. But this uncertainty about the importance of vision hasn't discouraged researchers who believe that the right hemisphere and its emotionalism are driving baby-holding to the left. Some of them think the connection may be based not so much on vision as on sound.

In 1996, communications scientists Jechil Sieratzki and Bencie Woll in London, England, suggested that sound was likely much more important than sight. They argued that mothers across cultures make the same kinds of sounds to their babies, that the sound of a mother's voice has been shown to be more important to an infant than her heartbeat (another nail in the coffin in the heartbeat argument for left-side holding) and that the right hemisphere adjusts the emotional content of those maternal sounds. So again, positioning might be critical: baby's voice is heard preferentially, because of brain wiring, by the right hemisphere; the right hemisphere in turn reads the emotion being expressed instantly, and so the bond between the two is strengthened. If the baby is on the right side, as Sieratzki and Woll wrote, "the lullaby will not sound the same."

Unfortunately, gathering evidence to support this idea has been as difficult as confirming the idea that the connection between hemispheres might be a visual one. Oliver Turnbull of the University of Wales and Heather Bryson of Aberdeen University tested undergraduate female students for two things: their tendency to rely on their right hemisphere to decode the emotional message in a language they were unfamiliar with, and their baby-holding preference. You'd expect that left-holders would be strongly right-hemisphere reliant if baby-holding was determined by the perception of emotional sounds, but Turnbull and Bryson found no such link. They did confirm that three-quarters of the students preferred to hold a doll on the left side, but there was no association between that preference and a left-ear (right-hemisphere) advantage for detecting emotion. Turnbull and Bryson wondered if perhaps the left-side preference might actually be driven by different senses (sight, sound or even touch) in different people, but to date the experiments that would support that idea have not been done.

That is where the puzzle stands today. Even if the left-side preference has something to do with the specialization of the right hemisphere for emotion, something strange is going on here. Remember that the studies of art and sculpture revealed that there seemed to be substantial changes over time—from one century to the next in Europe, preferences changed from left to right and back again. Such reversals would be very difficult to explain on the basis of brain hemisphere differences. You would have to argue that what appear to be hard-wired differences (even the great apes have them) can change over a relatively short time. Also, the examination of Central and South American art showed that regional differences existed at the same time. If you then take into consideration one additional study revealing that in Madagascar, women prefer to hold their babies on the right, not the left, the puzzle deepens even further.

If the secret lies in the brain, then explanations will have to be found for these contradictions that seem to have existed over time, and from place to place. That won't be easy, for however plastic and adaptable the brain is, it would take some powerful influence—as yet unidentified or even unimagined—to switch hemispheric specializations. Such switching does occur in some, but by no means all, left-handed people, but there's no evidence that baby-holding follows handedness; in fact, most left-handers hold babies on the left too.

It might be that we're still not close to an answer here. I have to admit that when I heard the brain-hemisphere theory for the first time, I found it hard to believe. After all, the advantage gained by having the "emotional" right hemisphere the one on duty must be marginal at best. It doesn't take a right hemisphere to hear the tone of distress in a baby's cry or see the happiness in a smile. Maybe the scientists are making too much of it. But if it isn't the brain, then what is it?

Counting
Coots

The American coot is not a particularly spectacular bird. Mostly black, it skulks around in marshes and would likely escape the notice of all but birdwatchers, and even those would likely spend little time oohing and aahing at the bird but would simply tick it off the list and move on. But the American coot has demonstrated an amazing ability, one that has rarely, if ever, been unambiguously demonstrated in animals in the wild: it can count.

Counting has been forced on the female coot by the pressure of evolution. Each female coot produces an average clutch of eight eggs, but apparently that's not enough, because those same females also have an irresistible urge to lay a few extra eggs in the nests of other female coots. Perhaps a few is understating the case: Bruce Lyon of the University of California at Santa Cruz found that female coots in British Columbia generally had three eggs in their nests that weren't their own—and that was in addition to their own eight. This is reminiscent of the nasty habit cowbirds have of laying their eggs in the nests of other birds, except that cowbirds don't even bother to build their own nests, and prey only on other species. In this case, it's coot against coot, with the bird that lays the most eggs winning the age-old battle to promote its genes into the next generation.

Actually it's a little more complicated than simply laying the most eggs, because those eggs have to be incubated to hatching by the female resident in the nest, and for that to happen, she has to accept the eggs and treat them as her own. Acceptance is a huge disadvantage to the female, who is not the least interested in raising cootlets other than her own. It's a harsh life: Lyon's study showed that about half of all chicks hatched failed to survive, and that each time a female successfully incubated a foreign egg, she lost one of her own. So female coots have evolved strategies for detecting and eliminating eggs that don't belong to them.

This is trickier than it might sound. Most coot eggs look pretty much alike—although from one clutch to the other there are subtle differences in the patterning of spots and background colour—and so the females have to be pretty sharp to spot the intruders. It's clear that they can do it, though, because females somehow recognize and remove the eggs laid by others. They don't always remove a foreign egg: sometimes, a female will simply push it out to the edge of the clutch. Such a female is called an "acceptor." Lyon suspects this is done when the foreign egg is so close in appearance to the host female's eggs that she can't be absolutely sure it isn't one of hers. Rather than removing it completely and running the risk of killing one of her own, she pushes it out to the edge where it will eventually hatch, but later than the others that are unambiguously hers. If it turns out she was wrong and it is one of hers, it might still hatch and survive, but it's less likely to do so.

The most fascinating part of this research is that the females control their egg-laying by estimating the number of eggs in the nest—when there are enough, they stop laying. Those coots who allow foreign eggs to remain in the nest, the "acceptors," shut down their own egg-laying early to ensure that their total number of eggs in the nest is what it should be. But those who toss foreign eggs out ("rejectors") do something different: they keep laying their own eggs until the nest is unusually full, and only then do they kick out the intruders. To do that, they have to be able to count the number of their own eggs versus the number of foreign eggs, and on that basis then lay as many more of their own as necessary. (If they were using some other strategy, such as estimating the amount of space left in the nest, both acceptors and rejectors would stop laying at the same time—but the key difference is that rejectors are counting just the eggs they're eventually going to keep, not the total they're looking at.)

It's even trickier than it sounds, because the female coot makes the decision to stop long before she actually produces her last egg—there's a delay of several days. So even though she might end up with eight of her own eggs, she makes that decision when she has only three or four: she turns off the egg tap at that point, but another four are already on their way. Nonetheless, it is clear that she is counting, at least up to four and maybe more.

If you're wondering what this could possibly have to do with the science of your life, just bear with me, and let me compare these coots with other birds and animals. An old naturalists' folk tale goes something like this (I've taken this version from Stanislas Dehaene's book *The Number Sense*): "A nobleman wanted to shoot down a crow that had built its nest atop a tower on his domain. However, when he approached the tower, the bird flew out of gun range, and waited until the man departed. As soon as he left, it returned to its nest. The man decided to ask a neighbour for help. The two hunters entered the tower together, and later only one of them came out. But the crow did not fall into this trap and carefully waited for the second man to come out before returning. Neither did three, then four, then five men fool the clever bird. Each time the crow would wait until all the hunters had departed. Eventually the hunters came as a party of six. When five of them had left the tower, the bird, not so numerate after all, confidently came back, and was shot down by the sixth hunter."

If only the crow had been a coot, it might have survived! This fable contains some truths that have been well attested by experiment. One, many different kinds of birds and animals can count, among them creatures whose intelligence would never impress us. Second, their numerical abilities are limited: if the crow in this story could indeed count accurately to five, it would have been an exceptional bird. But what is really amazing about the numerical abilities of animals is that it appears human infants share those abilities.

The animal experiments are straightforward. Pigeons are good experimental animals for this purpose: they can easily be taught to peck at a button to receive a food pellet. Once they've mastered that, they can be taught to press three times, or five times, or even twenty-four times. In fact pigeons are capable of a degree of number discrimination you might think was impossible for the not-much-loved "flying rat." They can be taught that pecking on one button forty-five times yields food but a different button requires fifty pecks. Then the bird is placed in front of three buttons, the forty-five-peck button on one side, the fifty-peck button on the other, and a third button in the middle. The third button lights up and the pigeon responds by starting to peck. When the bird reaches either forty-five or fifty pecks, the light is turned off. To get the food treat, the bird must remember whether it stopped at forty-five or fifty, then turn to the appropriate button and peck it, which it does with regularity!

Other experiments have shown that raccoons can learn to pick a box with three grapes (and three grapes only) inside, birds can choose the fifth seed that they find and rats apparently have some sort of abstract sense of number in their little brains, because they understand that (or at least push the right buttons when asked if) two flashes of light are the same as two sounds. In fact, once trained to push a lever upon hearing two tones or seeing two flashes, they will push that lever if a tone and a flash occur simultaneously. We're not talking here about brainiacs like the great apes or dolphins. These are—in some cases literally—birdbrains. But they have a good sense of numbers.

What about humans? For a long time the idea that infants would have any clue at all about numbers was not taken seriously; in fact it wasn't really taken at all. There were experiments and a whole school of thought that declared that learning numbers was a gradual, time-consuming process. But that notion began to be overturned in the early 1980s. A psychologist at the University of Pennsylvania, Prentice Starkey, directed a set of experiments that showed for the first time that very young babies could count.

The experiments were tricky to set up, because obviously it's hard to ask babies what they are thinking. Starkey used the technique of surprise: young babies will pay attention to things as long as they're not bored by them. The instant boredom sets in, their attention turns elsewhere. (They are born television watchers.) So Starkey had babies between sixteen and thirty weeks old sit on their mothers' laps facing a screen on which a series of slides of dots or actual objects were projected. A video camera recorded how long the baby looked at each slide. When the babies were presented with slide after slide showing two black dots, the time they spent looking at each declined steadily. But as soon as three dots appeared, the babies held their gaze longer: 2.5 seconds compared with 1.9 seconds. These experiments were later repeated with much younger babies, even those just a few days old, and varied, for instance, by having the babies look not just at dots but at slides of objects, some near, some far, some large, some small, in an attempt to eliminate the possibility that the babies were being cued by some difference in the slides other than numbers.

Two other kinds of experiments have broadened the picture of what very young children know—or don't know—about numbers. Prentice Starkey was again involved in one. He and his team positioned babies in front of two slide projectors, one of which showed two objects, the other three. A loudspeaker placed between the speakers sounded drumbeats. After the babies settled in, they spent more time looking at the slide of two objects when the speaker was producing two drumbeats, and longer at the slide of three when there were three beats. Starkey and his colleagues argued that the babies were linking the drumbeats and the objects on the slides, showing that they had an abstract sense of number: to them, two meant two of anything, not just two objects on a screen.

The third variation on this theme was the most dramatic, and the first of its kind was performed by Karen Wynn of Yale University in 1992. Wynn used the same technique of monitoring babies' interest and surprise as a measure of what they understand, but in this case the set-up was a little more elaborate. Wynn wanted to know not just if babies knew their numbers but if they could work with them. Could they add and subtract? She began with the well-known fact that even small babies react strongly when something that appears to be impossible or magical happens. So if an object seems suspended in mid-air, they are transfixed. If something is hidden behind a screen, but then isn't there when the screen is removed, they can't believe it, and stare intently as if expecting the object to return.

Here's how the experiment worked. Four- and five-month-old babies were seated in front of a miniature puppet theatre. They first saw the hand (of the experimenter) place a toy Mickey Mouse on stage. Then a screen came up, hiding Mickey. The hand then reappeared with a second Mickey, put it behind the screen as well and then (and this is important) could be seen leaving empty. So to this point, if the babies could follow the math, it had been one Mickey plus one Mickey. But sometimes when the screen was removed, there was only one Mickey left—the other had been removed through a secret trap door. The babies reacted as if something forbidden had just happened. They stared at this unexpected outcome, looking a full second longer than they did if the screen revealed the expected two Mickeys. The babies showed even more surprise at the reverse experiment: the magical and impossible $2 - 1 = 2$.

These experiments showed that human babies are about on a par with animals when it comes to simple math. (In fact Wynn's experimental design has been duplicated with rhesus monkeys doing the counting and eggplants standing in for Mickey.) But they aren't without controversy. From the beginning, critics have argued that babies might be detecting features of these displays other than simply numbers.

If the dots were black on a white background, maybe the babies were looking at the total amount of black on the slide and not the actual number. Or could they have been paying attention to the brightness of the slide, which would obviously change if it were obscured by more black dots? Some critics have gone so far as to suggest that babies might be attending to the "contour length" of the displays—that is, the total length of the circumferences of the dots. Innumerable experiments have been designed and redesigned to take these possible flaws into account.

Karen Wynn's addition and subtraction experiments have also received intense scrutiny. One suggestion has been that, given the design of the experiments, babies spend much more time seeing just one Mickey than two. (The experiment always starts with one, and sometimes ends with one as well.) If—and I suspect this is a big if—the baby participants are overwhelmed by the complexity of the experiment, they might simply have preferred to look at the situation they were most familiar and comfortable with—one Mickey.

Another possibility in the 1 + 1 = 1 version is that these babies are simply expecting more objects, and are surprised when the number of Mickeys remains the same. Wynn dispelled that objection by having the babies see two alternative outcomes: 1 + 1 = 2 and 1 + 1 = 3. If babies were simply expecting more objects, then they should have had no preference between these two, but in fact they were much more interested in the impossible 1 + 1 = 3 version, again suggesting that they could count and anticipate what the outcome should have been.

Sometimes the efforts to dispel doubts can be elaborate. Karen Wynn designed an experiment that showed babies moving swarms of black circles. The assumption was that circles that moved together, like bees in a swarm, would be interpreted by the babies as a single "thing." In the experiment, some babies got used to seeing four groups of two circles each, while others watched two groups of four circles each. Note that the total number of circles (and every feature of them) was the same, but the number of "things" was either two or four. Wynn was confident that once the images were switched, the babies would perk up, because they would be seeing different numbers of objects, even though the numbers of circles that made up those objects were the same. And she was right. I doubt this latest experiment will resolve the controversy over infant numeracy, but to my mind the evidence that very young babies have some numerical knowledge is pretty convincing.

There are some intriguing spin-offs from this research. One, the knowledge of numbers shows up so early in life that it is likely that it is innate. Some scientists have even argued for a kind of number module in the brain, just as others have claimed there is a language module. Some sort of innate module would extend pretty far back in time, because presumably those animals that share with human babies some of these elementary abilities to use and manipulate numbers have a version of this module themselves.

One would have to travel far back in evolutionary time to find the common ancestor of the human infant and the calculating pigeon, but they share not only these numerical strengths, but weaknesses as well. Both pigeons and humans—even adult humans—have trouble discriminating between numbers as they get larger, or closer together. For instance, if you are asked to tell, as fast as you can, which of a pair of numbers is larger, you will be able to do that much faster if the numbers are 3 and 6 than if they are 33 and 36. Also, the further apart they are, the easier it is. Those pigeons that can tell the difference between 45 and 50 have a much tougher time if the numbers are 49 and 50. These are numerical limitations that stay with us throughout our lives, and even extensive training seems incapable of changing the situation much.

Here's something else. Babies are pretty good with the numbers 1, 2 and 3, but beyond that they start to have trouble. We adults are no different. Ask adults to identify the number of dots appearing on a screen, and they are extremely fast as long as there are fewer than three. But once the number of dots rises to four, the response times start to soar: for one, two and three dots, the average time to figure out how many there are is six-tenths of a second, but that grows to more than eight-tenths with four, and to more than a second with five. At the same time, the number of errors rises too.

Probably one shouldn't read too much into this, but it might be possible that we can count the first three numbers with such ease because we can rely on an innate number module that deals very well with those numbers but not so well with anything higher. It might even have had cultural impacts: Stanislaus Dehaene and others have pointed out that several numbering systems around the world, including Roman numerals, use straightforward designs for the first three numbers (I, II, III) but then depart radically beyond that (IV). All I know is that I distinctly remember reading many years ago that some far-away culture (no doubt in those days it would have been labelled "primitive") had four numbers: one, two, three and "many." It wasn't that they were ignorant of higher mathematics; they were just giving voice to their number module.

One final thought concerns the number 0, a number that sits very close to the 1, 2 and 3 that these experiments focus on, and yet a number that is in some ways completely different. Karen Wynn's addition and subtraction experiments turned up something weird when she presented infants with $1 - 1 = 1$. (To do this a Mickey has to be snuck back onto the stage.) When this happens, the infants evince no surprise at all. It was as if, in contrast to all the other experiments, they had no expectation about the outcome of $1 - 1$ at all. That may seem odd to us, but there is something special and unusual about zero. For one thing, it shows up relatively recently in the history of mathematics—the Babylonians were using numbers for 1500 years without bothering to invent a zero, and even when zero finally took its place among numbering systems in various cultures, it was used as much as a designation of an empty space as a number in its own right.

One wouldn't want to stretch this analogy too far, but it appears as if the delayed recognition of zero as a number in history is reflected by the same delay in childhood. Karen Wynn's infants don't seem to know what it is, and nor do children much older: preschoolers who know that zero means nothing will still argue that the smallest number is 1. The question is, would a coot know that zero is a number?

Looking
for the
Unattractive Man

The whine of a mosquito on a cottage evening is as memorably Canadian as the call of the loon. Although I've read a lot of popular accounts and seen many drawings of the female mosquito's proboscis and how its stylets pierce your skin in search of an available blood vessel, the prelude to that nasty moment is usually ignored. How did the female find you in that dark room? Did she see you, smell you, hear you or sense your presence in some other way? Would it be possible to design better repellants if we knew exactly what these chemicals do to discourage female mosquitoes on the hunt?

Simply watching the dreaded insects in flight provides some clues as to how they find us. The following description of mosquito behaviour is an unusually evocative one for a scientific journal. It's from an article by Anthony Brown in the 1966 *Journal of the American Medical Association:* "When a man is standing in mosquito-infested territory in a light breeze, the mosquitoes will pass him by, but those which enter the downwind cone of his emanations will turn and work upwind to him from as much as 30 feet away."

Being in the downwind cone of someone's "emanations" can, at least in our species, be dramatically unpleasant, but obviously female mosquitoes are, well, different. "When a man is standing in a closed room, the mosquitoes in flight alongside ignore him, but let him lie down and he will activate all those which experience the ascending convection currents above him."

The references to a man in these scenes isn't simply sixties sexism—men are more prone to being bitten than women. (It's also true that adults are preyed upon more than children.) But words like "currents" and "emanations" make it sound as though odours or vapours—not sights or sounds—play the crucial role in attracting mosquitoes. If that's true, then maybe repellants somehow interfere with the mosquito's ability to sense those odours and figure out where the source is.

But we're getting ahead of ourselves. It may be true that odours are the key player here, but mosquitoes can and do see us. (There's no evidence at all that our sounds attract mosquitoes, unless we mimic their ultra-annoying whine, and even then it would only be the non-biting males that would home in on us, looking not to bite but to mate.) Vision is not only important, it might even be the way that female mosquitoes first notice us. The darker the object, the more attractive: apparently mosquitoes' eyes are not sensitive to the colours in the middle of the rainbow, from orange to blue-green. However, colours further out toward the end of the spectrum, red in one direction and violet in the other, probably look black to hunting mosquitoes and do arouse their interest. If a dark object is moving, that's even better. And if the object produces a flicker effect by moving an alternating black-and-white surface, especially striped or checkerboard, that's the best of all. "If you're dressed up like a referee, Don't go into the woods today . . ."

However, most of us know that even if we wear white clothes in mosquito-infested woods, we're still going to get bitten, and probably bitten a lot. Can we get even better clues about what attracts mosquitoes by finding people who, clothing aside, either do or do not turn them on?

The answer to that question is a qualified yes: there are people who are more or less attractive to mosquitoes, but figuring out why—that's another thing. In one memorable effort decades ago, a team in California tested 838 people for their attractiveness to mosquitoes. Each person had a small mosquito cage (a plastic cylinder with mesh nylon net on the ends) pressed against the forearm for three minutes. There were only four mosquitoes in the cage at any one time, but nonetheless in the first run of tests only 17 out of 838 people were unbitten. When these lucky seventeen were retested, only one remained unbitten. This man was tested nine more times, and he was bitten only three times. Bottle that guy! He was truly the "unattractive man" of the title of this chapter, but unfortunately, even though the researchers assured us in their report that "exhaustive studies" of this man were performed, no clue to his unattractiveness emerged.

That's pretty well where the story stands today: there is no question that some people are more prone to being bitten than others, but it's not clear why. Considering that hundreds of chemical compounds are continuously being released from the typical human body, it wouldn't be surprising if different mixes of them sent different signals to mosquitoes, but teasing those chemicals out is the hard part. One twenty-one-year-old man studied by this same team in the 1960s was incapable of sweating, and attracted fewer bites, but that's a pretty high price to pay for freedom from mosquitoes—being unable to sweat is a serious disability. However, this was a dramatic illustration of something that other investigators have found since: sweat plays some role in attracting mosquitoes, and presumably some people's sweat smells sweeter than others'.

In the mid-1990s an experiment in Tanzania confirmed the idea that people differ in their mosquito rating. Three volunteers slept for nine nights in tents that were fitted with a mosquito *entrances* and exit traps. (They could have saved some of the effort expended in making entrances by inviting me to put up their tents; that would have guaranteed the presence of hordes of mosquitoes.) The volunteers switched tents over the course of the nine nights, taking their bedclothes with them to ensure that any differences in bite frequency couldn't be attributed to lingering odours. One of the three turned out to be significantly less attractive to three species of mosquitoes that entered the tents.

But it's not just the difference of one person to the next: one Japanese experiment found that some body parts are more attractive than others. They identified feet as the preferred target for mosquitoes, followed by hands, then the face. Why this order of preference, no one knows. There's even evidence that mosquitoes will land much more often on people after they've consumed a beer; more bad news for Canadians.

Researchers have been able to identify some of the "emanations" from humans that attract mosquitoes, but it hasn't been easy. A combination of recordings from single nerve cells in the mosquito's antennae (a feat in itself) and detailed pictures taken of those same antennae by the electron microscope have established that about 90 percent of the nerves in the antennae are devoted to detecting airborne chemicals, and that carbon dioxide, warmth and lactic acid are all powerful attractors for mosquitoes. All make sense, of course: we are warm, we exhale carbon dioxide and we leak lactic acid in our sweat. What is more interesting about this trio of turn-ons is the clever apparatus the mosquito uses to detect them, a set of detectors wired to ensure that she gets to her target efficiently and safely.

Our breath is about 4 percent carbon dioxide, or CO_2. That is a hundred times more concentrated than in the air around us, so it is not surprising that an insect equipped with CO_2 detectors could pick us out of the gaseous background. Still, it's not exactly the yellow brick road: the carbon dioxide plume that emerges from our mouths is immediately caught by the breeze and fragmented into wisps and filaments of gas, as well as being instantly diluted. But the insect's CO_2 receptors are equal to the task: they are perfectly tuned to the gas concentrations they might encounter, their sensitivity topping out at that 4 percent level, but at the same time capable of sensing changes in concentration as small as 0.01 percent. Not only that: they are pulse-sensitive. Mosquitoes in a wind tunnel exposed to a steady stream of carbon dioxide, quite unlike anything they're likely to encounter in the real world, choose not to fly upwind. Yet mosquitoes that are standing still can be induced to take off by a puff of carbon dioxide.

The mosquito's CO_2 detectors are not on the antennae but instead are on stubby little projections from the head called "capitate pegs." Their role is easily demonstrated: amputate them and the mosquito is unaware of the gas. In a more sophisticated vein, recordings from the nerves exiting these pegs show that they are activated by whiffs of CO_2. Incidentally, those nerves have an architecture that seems to be more complex than it has to be for the job at hand, a bit of over-engineering that isn't yet completely understood. It might have something to do with coordinating the carbon dioxide signal with others the mosquito is attending to, because her sensitivity to other odours (such as lactic acid) is conditional on the carbon dioxide being there at the same time. That coordination is a little tricky, because it must involve matching signals gathered from the antennae with the CO_2 signals recorded at the pegs, a little like matching the sight of a grilled cheese sandwich with its odour.

Male mosquitoes are attracted by carbon dioxide, even though they don't bite. The best guess here is that the males are simply looking in the likeliest place for females (as they always do) and females will be found wherever there are mammals exhaling carbon dioxide.

Those same mammals are of course warm, and the female mosquito's antennae are studded—especially at the tip—with temperature receptors of high sensitivity. These receptors respond quickly to changes of a mere five one-hundredths of a degree Celsius by firing off volleys of nerve impulses, and lab studies have shown that such minute traces of warmth can be picked up in the air currents rising from a rabbit that is two metres away from the mosquito. At the same time, these temperature detectors react only mildly to the slow changes of temperature the insect would encounter as it flies from place to place. It appears as if the mosquito's temperature sensitivity is higher in moist air, but there's no convincing evidence yet that they actually have sensors for humidity.

On the other hand, there is no doubt that they home in on lactic acid, using yet another specialized set of receptors on the antennae. As is the case with carbon dioxide, those receptors are tuned to the range of concentrations of lactic acid that escape from human skin (and only from human skin). It's a safe bet that when a female mosquito is pursuing you, her lactic acid receptors are in a state of high alert, although that isn't always the case. There are times in a female's life when there is no point for her to continue pursuing prey, such as when she's just hatched, or already full of blood, or already incubating a set of eggs. At these times those same lactic acid receptors that generate volleys of nerve impulses are silent, probably dampened by hormones circulating in the female's blood. It has even been possible to perform the mosquito equivalent of a blood transfusion from such a mosquito to a hungry one, and the recipient immediately stops looking for prey.

This control of the female's lust for blood is another example of the sophistication of the mosquito's behaviour, sophisticated at least for an unthinking, chitinous robot. She doesn't "decide" whether or not to attack you—she does what she's wired to do. A mosquito attack may be annoying for us, but it's potentially fatal for the insect. So just as her approach to you is guided by nerves firing in response to the presence of chemicals in the air, her decision to avoid you is also neurochemical.

When a mosquito is full of blood (5 microlitres' worth, a mere smear on the skin after it's slapped), stretch sensors in her distended abdomen send a message to the brain to stop feeding, and she pulls out her needle-like mouthparts and takes off. The efficiency of this system has been dramatically demonstrated by cutting the nerves that carry the message from the abdomen: the mosquito will continue to ingest blood until she explodes. Being full of blood not only prompts her to take off but also inhibits her from either returning to you for more or seeking another victim. Although her initial avoidance is dictated by her nervous system, eventually hormones are released to sustain that avoidance.

Warmth, carbon dioxide and lactic acid are probably the three most significant attractants for mosquitoes (although how they relate to mosquitoes' apparent preference for feet, hands and face—in that order—isn't clear) but they are not the whole story. While you can attract mosquitoes with an artificial mix of warm moist air containing carbon dioxide and lactic acid, those mosquitoes will not probe—that is, they won't land and start dabbing at the surface looking for blood vessels. This is just one example of evidence that there must be other chemicals that mosquitoes are attracted to, chemicals that together provide the complete human attractant package: there's no mixture yet that has been concocted in the lab that can attract mosquitoes like a warm, somewhat sweaty human being. They are being identified one by one, though. One clever scientist noted that the preference of one species of mosquito for human feet suggested—by analogy—that the same insect might be attracted to cheese, Limburger cheese in particular. It was, probably because there are similar species of bacteria acting to produce the odour in each.

Knowing what we know about what attracts females, can the action of repellants be explained? There are several possibilities. A repellant could coat the antennae and prevent odours—whatever they are—from coming in contact. That might even require the repellant to plug directly into specific odour receptors on the antennae. On the other hand repellants could do the opposite and confuse mosquitoes by overstimulating them instead. Or they might have nothing to do with the mosquito itself: they could work by preventing odours from escaping your skin, leaving the mosquitoes with nothing to sense.

There are hints from simple observations of how mosquitoes behave in the presence of repellants. In the 1960s and 1970s scientists watched as female mosquitoes flew in and out of airstreams that contained, as they put it, "host-related stimuli." Females didn't seem to react when they flew into such air, but did turn to re-enter if they happened to fly out of the stream. By contrast, if a mosquito entered a stream of air containing a repellant, she either turned away when entering the stream or failed to re-enter it upon leaving. So somehow the repellant was changing the mosquito's behaviour, although exactly how could never be figured out from these simple observations.

But that was the 1960s, and science has moved on. For one repellant, there's now solid, clear-cut evidence that it messes up one of the mosquito's primary host-detecting systems. That repellant is DEET. (Just a note about the name: DEET was originally called N,N-diethyl-m-toluamide, which was at least a close approximation to the initials. However, it's had a name change and is now called N,N-diethyl-3-methylbenzamide, which bears no resemblance to "DEET" at all but is the same stuff: the best mosquito repellant—by far—that anyone has ever come up with.)

There never was really any doubt about DEET's pre-eminence as repellant par excellence, but its status was reaffirmed in the summer of 2002, when a study in the *Journal of the American Medical Association* compared DEET with a suite of pretenders and found that it outstripped them all when it came to comprehensive and long-term protection against mosquito bites. It all boiled down to just a few numbers: the highest concentration of DEET protected for about five hours, whereas its closest competitor, a soybean oil repellant, worked for only about two. Citronella was so ineffective it practically attracted mosquitoes. One eucalyptus oil–based repellant came to the attention of the researchers too late to be thoroughly examined, but it did, on preliminary testing, seem to be pretty good. There are also other repellants not tested in this study that have worked well in different settings with different mosquito species. Any of them might turn out to rival DEET, but for now N,N-diethyl-3-methylbenzamide is king.

These findings haven't been embraced with the enthusiasm you'd expect, even in the West Nile era, because most people view DEET with great suspicion, usually because it has dissolved their sunglasses or the crystal of their watch. It is, as they say, a potent *plasticizer,* and that leads to the suspicion that it might just do that to one's skin or internal organs. It does dissolve right into and through your skin, it does irritate mucous membranes such as your eyes and lips (if you're unfortunate enough to put it there), and there have been serious reactions to DEET reported in the medical literature. But those reactions add up to fewer than fifty over forty years and eight billion doses. Three-quarters of those fifty medical complications resolved without any lasting effects. I'm not an apologist for DEET: it has been responsible for some dreadful, even fatal side effects over the years, most of those in young children. But the evidence suggests that 10 percent solutions of DEET, carefully applied, are not a significant health risk for children. Adults shouldn't use anything over a 30 percent solution themselves, but that is no real hardship, even in the deep woods. Raising the concentration to 50 percent from 30 percent extends the repellant effect only minimally.

That repellant effect is dramatic. One study in Alaska, where the air was literally thick with mosquitoes, showed that DEET on the skin reduced the number of mosquito bites from a maximum of a stunning, insanity-producing 3360 bites an hour to 1 (!) an hour. There is no doubt that it works, and there's solid understanding why: DEET appears to turn off the lactic acid receptors on the mosquito's antennae. The nerves connected to those receptors simply don't fire when the receptor is exposed to lactic acid. Apparently it doesn't matter that carbon dioxide and warm moist air, the other major attractants, are present. The shutdown of the lactic acid receptors blinds the female mosquito to our presence.

DEET's effect can even be seen in the way the mosquito behaves. When she flies into plumes of air carrying no attractive odours, she flies erratically, turning this way and that, sometimes as much as 90 degrees or more. By contrast, the same mosquito in a plume of lactic acid maintains her bearing roughly upwind, turning no more than 45 degrees in either direction. When faced with an airborne mixture of attractant and DEET, she reverts to her aimless ways, apparently casting about for some scent of a target. It's not that she's turned off by DEET—it simply makes her unaware.

Ultimately the molecular biologists might be able to show exactly how DEET and those lactic acid receptors interact. If there is some sort of lock-and-key fit, it might then be possible to design even more effective, longer-lasting and safer repellants. But for now we are a long, long way from the total picture of how mosquitoes find us, and how repellants interfere with those mechanisms.

Echolocation—
Our Sixth Sense?

Echolocation is the wonderful sense bats and dolphins use to navigate without sight. We know most about bats' echolocation: they can find a flying insect in the dark, judge even its size and texture, and then home in on it by emitting ultrasonic squeaks and listening for the echoes that bounce off the unfortunate prey. It is a sense exquisitely attuned to the bat's everyday—or everynight—life, but it has nothing to do with us. The philosopher Thomas Nagel once wrote an article called "What Is It Like to Be a Bat?" in which he argued that we could never imagine what it was like to be a bat because theirs is a world constructed by echolocation. But is echolocation beyond our wildest dreams? Maybe not.

There is a small group of researchers today who suggest that we may be capable of using echolocation, or, more dramatically, that we may actually be using it and just not know it. Even though this would qualify as a sixth sense, it wouldn't be surprising if we weren't aware of it—we are so reliant on vision to navigate our way around that the relatively small effects of reflected sound would likely be swamped by what our eyes are telling us. But people who cannot see might be able to take this sense and elevate it to something of great significance.

The idea that humans might be able to echolocate has a long history—there are anecdotal reports going back more than two hundred years of blind people being able to sense the position and distance of objects in their path. But the first serious and detailed experiments were performed in the early 1940s at Cornell University. In the first of those experiments, published in 1944, two blind and two sighted but blindfolded individuals were required to approach a wall and make two signals: one when they thought they had first detected the presence of the wall and a second when they were sure they had moved as close as possible to it without colliding with it. One of the two blind participants was incredibly good at both: he could apparently sense the presence of the wall from distances of 8 metres or even more, and he could routinely approach to within 15 centimetres. But even the sighted pair became pretty good at finding the wall, and though their numbers weren't as impressive, their achievement was: after all, it was expected that only someone who had been blind for many years would have developed any "sixth" sense to its maximum.

But exactly what that sense might be wasn't clear. The exceptionally talented blind person in this study was certain that his awareness of the wall had to do with a kind of "facial vision" that he felt as he closed in on it. This wasn't a new idea. In the nineteenth century there had been sporadic attempts to demonstrate the ability of blind people to "find" obstacles in their path, and more than once the suggestion had been made that somehow unspecified sensors on the face detected changes in air pressure or air currents stirred up by the obstacle. Some even went so far as to claim that the "distance sense" was strongest around the eyes and ears and weakest around the mouth and lips. But these remained suggestions only, because no one had been able to demonstrate that any sort of facial vision—a sense that the blind had and sighted people didn't—existed.

The only reasonable alternative explanation was that somehow these people were able to use sound to detect objects in their path, something that other investigators in the nineteenth century had strongly suspected. The idea is that sounds reflect from surfaces, producing mini-echoes, and that those echoes somehow change the closer the approach. The Cornell experiments were designed to distinguish between the two possibilities. The experimenters, Milton Cotzin and Karl Dallenbach, plotted out a remarkably painstaking and exhaustive set of experiments to determine exactly what was necessary for participants to be able to find the wall in front of them. These included walking with shoes on a bare floor, walking barefoot on carpets, wearing elaborate felt veils to block any breezes or pressure of any kind from hitting the face, wearing earplugs to diminish the intensity of sound and, most remarkably, sitting in an isolated room wearing headphones and listening to the sounds made by one of the experimenters as he walked toward the wall carrying a microphone.

The results were clear-cut: any technique used to reduce the amount of sound that the subjects could hear diminished their ability to find the wall; any technique that somehow interfered with "facial vision" had no effect on their abilities. Even the blind subject who had been absolutely sure that he felt pressure on his face gradually came to admit that he had been wrong. His first clues came when his performance declined dramatically if he approached the wall wearing only socks on a double thickness of carpet. In that case he felt that the "pressure sensations do not feel anywhere nearly as strong as before." He finally had to give in completely when he wore a veil while approaching the wall and realized that he was desperately listening for any clue to his whereabouts: "I find myself scraping my stockings on the carpet in an endeavour to make a little noise." By the end of the experiments he acknowledged that sound must have been what he was using.

Karl Dallenbach, the lead investigator in this experiment, went on to perform others: one to determine what sort of sound frequencies were best for echolocation (the higher-pitched, the better) and another to determine whether abandoning the somewhat controlled indoor environment used in the first experiment for the outdoors would have any impact on people's abilities to echolocate (it didn't).

The experiment designed to pinpoint the best frequencies of sound produced some unusual testimony from the participants, some of whom were the same as in the 1944 experiment. This experiment differed from the first in that the subjects didn't walk but sat in a chair in a separate room while a robotic loudspeaker, moving silently along wires suspended from the ceiling, approached the wall while emitting a variety of sounds. The subjects listened through headphones, hoping to pick up clues that the wall was getting closer. They did amazingly well, considering the artificiality of the experience. When a form of white noise (a sound combining all the audible frequencies) was beamed at the wall, both the blind and the sighted subjects could detect the wall when it was a few metres away, and could approach to something less than 30 centimetres before asking the experimenter to stop the forward movement of the loudspeaker. They were also tested with pure tones, like those produced by a tuning fork, and in this case their behaviour was much stranger. Either they collided with the wall or, once having signalled their initial awareness of the wall, they were unwilling to approach any closer. Somehow pure tones provided only one piece of information—that the wall was there—but nothing more. However, one pure tone produced startling effects, and this one was extremely high-pitched: 10,000 cycles per second, equivalent to a note more than five octaves above middle C, well above the highest note on a piano keyboard.

The subjects had weird experiences with this high-pitched tone: "with this tone there are fluctuations like ripples on water"; "The tone becomes more piercing and shrill when it nears the obstacle"; "it screams when near the obstacle." Despite these vivid experiences, they weren't sure exactly what it was about the tone that led them to detect the obstacle. Was it volume, pitch or both? More experiments determined that it was a change of pitch, and the experimenters decided that it must have something to do with the Doppler effect.

This effect is the dramatic change in pitch of a train whistle or a car engine at high revs when it rushes past. It happens because the waves of sound pile up as the train or car approaches (so producing a high frequency), then are spaced out as it rushes away, leading to a lower frequency. It works with these vehicles because they are travelling at a high enough velocity to produce a significant difference between the two sets of sound waves. But in most echolocation experiments, the only movement is that of people at walking speed, not by itself enough to produce the effect. Yet as Dallenbach pointed out, in this experiment both the sound source (the person's footsteps or the loudspeaker) and the receiver (the ear or the microphone) are moving toward the wall, so the effective speed is doubled. Even so, his calculations showed that the only pure tone they had used that could possibly yield a discernible change of pitch at that speed was the one that actually worked: 10,000 cycles per second. Of course we don't walk around listening to high-pitched pure tones in everyday life, but we do hear sounds, like footsteps, that are mixes of frequencies, some of which are sure to be in that upper range.

These experiments half a century ago seemed not to inspire much interest in human echolocation. It was a decade more

before another research report was published, this time by
Winthrop Kellogg at Florida State University. Kellogg tested the
ability of his subjects to judge the distance, size and texture of
objects that were placed before them while they were seated. The
two blind and two sighted but blindfolded subjects would have
been amused by the scene before them had they been able to see:
discs of quarter-inch plywood were attached to ropes, and the
ropes to bicycle wheels, so that the discs could be suddenly
shifted closer to or farther from the subjects' faces. Given that
there was no sound of footsteps to create echoes, the participants
in this case were allowed to make any noises they wanted. Some
snapped their fingers, some hissed, clicked their tongues or
whistled, but most spoke or sang. (All these sounds contain the
high frequencies that are likely to be most useful in tracking
echoes.) The blind subjects added a behaviour that dolphins use
when navigating in murky water: they moved their heads from
side to side, doing something Kellogg called "auditory scanning."
This was apparently an attempt to get a better fix on the distance
of the wooden disc by comparing the time of arrival of sound at
one ear with the other.

Kellogg showed that his blind subjects were very skilled not only
at determining the distance of the disc but also at telling the difference between discs made of different materials. One subject could
tell when the one-foot-diameter disc had been moved either closer
or farther away by four inches or so, and both could discriminate
between discs made of denim, wood and metal.

All these experiments involved blind people who had been using
echolocation (even if they didn't realize it) for years, or sighted
people who participated in both lengthy training and experimental
exercises and so had a chance to sharpen their latent echolocation
skills. But what about the rest of us? Do we use echolocation in any
way during a typical day? And if we do, are we aware of it?

Eric Schwitzgebel at the University of California Riverside argues that we probably do use it but that it's almost certain that we don't know it. After all, he says, even those participants in the experiments back in the 1940s weren't really sure what they were doing to sense the presence of an approaching wall, and the one who was best at it at first thought he was using "facial vision," a kind of pressure-sensing ability. Schwitzgebel points out that the aural experience of walking down a tiled hall is very different from that of stepping across the tiles on a bathroom floor: if one sound were substituted for the other, it would seem quite strange. Schwitzgebel also suggests that if you opened a door and started down a familiar hallway, failing to see that someone had left an overstuffed chair in the hall directly behind you, you might discover the chair by noticing a change in the sound of the hall.

You can experience for yourself, in a small way, the nature of echolocation by holding your hand in front of your face, about arm's length away, and then moving it slowly toward your face while you hiss. You'll notice that at some distance, the sound of your hissing will change, then continue to change as your hand moves ever closer. The change of sound you hear is not a simple one: it's been called "ripple noise pitch" and it results from the outgoing hissing sound interfering with the returning echo in an ever-changing way as your hand approaches.

Admittedly, hissing at your hand is a long way from being able to walk around and avoid collisions using only your ears, but at least it suggests that we can use echolocation to some degree all the time. However, when we do we're probably tracking much more than just the echoes of our voices or footsteps bouncing back from a wall in front of us. No matter where you are during the day, there are countless sources of sound around you, some near, some distant, some covering most of the spectrum of sound we can hear, many representing only a few frequencies. To be able to echolocate, the brain has to be able to select something relevant out of that mess of information. The experiments suggest that we're able to do that in a limited, laboratory-confined way. The next challenge is to show that even with our eyes wide open, we're still able to "see" our way around using our ears.

It's Time—
You Must
Wake Up Now

M ore than a hundred years ago the great American psychologist William James wrote: "All my life I have been struck by the accuracy with which I will wake at the same *exact minute* night after night and morning after morning, if only the habit fortuitously begins . . . After lying in bed a long time awake I suddenly rise without knowing the time, and for days and weeks together will do so at an identical minute by the clock, as if some inward physiological process caused the act by punctually running down."

James was describing a phenomenon that apparently is by no means unique to him: the ability to wake up at a predetermined time without an alarm clock. We all know people who claim to be able to do this—the question is, have they just fooled themselves into thinking they can do it (by believing that they really *were* awake when the alarm went, or by waking up eleven times in the night, the last time just before the alarm) or is there independent evidence that at least some people can do this?

One of the first attempts to investigate this thoroughly—and relatively objectively—was published by a British physician, Winslow Hall, in *The Journal of Mental Disease* in 1927 (although others had been researching it decades before). He had carried out exhaustive tests of his own ability to wake up, some of them as much as ten years before this publication. His accounts are touching to read because he admits that "bodily pain" of some kind usually wakened him several times a night; such awakenings—which would have nothing to do with a body clock counting off the hours until wakening—made the interpretation of his results more difficult. However, Hall reported that over one hundred trials, he woke up within fifteen minutes of the chosen time more than half the time, and at exactly the right time eighteen times. That is pretty impressive, but this report illustrates one of the frustrations of these studies: does the fact that Winslow Hall could wake up when he wanted to say anything at all about the rest of us? I'm not sure.

For one thing, Hall seemed to have an unusually acute sense of time, asleep or awake. In one experiment, he and three friends agreed that they would try to guess the time of day. At first, when signalled to do so, they gave themselves up to five seconds to guess what time it was, but later they decided to abandon the five seconds and simply shout out their time-guesses immediately. They actually did slightly better without the five seconds of reflection (or second-guessing): 45 percent of their guesses were within three minutes of the actual time, while 10 percent were exactly right. With the five-second delay, the results were 41 percent and 9 percent respectively.

Hall also tried imagining the face of his watch while it was still in his pocket. "On October 19, 1917, I imagined the hour hand approaching five. Then, on trying to see an image of the minute hand, I had a conviction that the time was twenty-two minutes to five; whereupon I could see, in imagination, the minute hand so placed. On looking at my watch I found the hands pointed to twenty-two minutes to five exactly." By his own count, Hall was able to guess within three minutes of the exact time more than 40 percent of the time, an accuracy comparable to the first experiment with his friends. However, Hall's experimental report, personable as it is, leaves out some of the procedural details that would shed useful light on the results: exactly how were the four experimenters "challenged" to guess the time, and what sort of cues to the time of day other than their internal clocks might have been available to them? Regardless, his results aren't out of step with studies that have been done since. As Hall himself wrote, "A time-sense exists."

One thing that is unique about Hall: he was unusually reliant on visualizing the face of his watch. He did it both in his day experiments and also before going to bed. In none of the experiments since has any technique like that been specified, so it's hard to say whether that would help others, or whether it was just Hall's way of keeping track of mental time. But imagery was obviously important to him. He reported in one of his successful awakenings (July 10, 1917, 4:30 A.M.) that he saw, in the middle of a dream, a hand-lettered sign, pink with black letters, with the word "TRUTH." It woke him up, but he wondered later whether an image of his watch might not have been more direct. There have been other suggestions that imagery plays an important role: in one study in Germany in the 1920s, some of the subjects reported that they heard, in the middle of a dream, the words, "It's time, you must wake up now."

Since Hall's report there have been many tests of the ability to wake up at a predetermined time. All show that some people can do it some of the time while others are absolutely hopeless, missing the appointed time by hours. None has identified any single individual who can wake up precisely as predicted every morning. Of course "precision" is relative. Getting up five minutes early or late usually isn't significant, and that sort of accuracy seems to be fairly common. However, there is still controversy about whether this ability really is as widespread as it might seem.

One skeptic, Harold Zepelin, pointed out that early studies, like Winslow Hall's, have to be viewed with caution because they lacked independent confirmation of the results by outside observers. Zepelin went on to argue that modern studies in sleep labs have revealed that most self-awakenings emerge from rapid eye movement (REM or dreaming) sleep, a time when waking is more likely anyway, and also a time of heightened awareness that might make it easier to remember the task at hand. A normal night's sleep includes several REM periods, each longer than the one before, making it more and more likely as the night goes on that REM sleep and waking up will coincide. So Zepelin was suggesting that what might look like waking on cue is simply a waking out of REM that might have happened anyway.

You'd think the idea of coincidence should be easy to test: just figure out how often people actually wake up at the designated time, and how likely it would be that they might wake up just by chance at that same moment. The problem is that every night's sleep is different, and even if you're the kind of person who wakes up several times during the night, it is extremely unlikely that you will wake up at the same time every night. You might hit the target time with one of those awakenings one night but not the next. But more important, decades' worth of studies suggest there are more on-time awakenings than you'd expect from coincidence. Something is going on here.

Other features of this phenomenon have emerged from the experiments. One is that people seem to know whether they have this ability or not—success has been shown to tally with subjects' expectations. Another is that motivation seems to play an important role. In one Israeli experiment seven subjects each tried on two successive nights to wake up at a predetermined time. Then the two best—those who woke up within ten minutes of the target time—spent another two weeks in the sleep lab, but their accuracy declined dramatically, to an average of nearly eighty minutes from target. I know from personal experience that the more important (or intimidating) the event, the more likely I am to wake up at or before the alarm; the Israeli experiment seems to bear that out.

A second intriguing feature is that the rates of success are about the same as those achieved when people try to estimate the passage of time while they're awake; that is, within a few minutes about 40 percent of the time. That does suggest that there is some kind of internal timekeeper that is consistent 24/7. In fact one of the most amazing experiments on the ability to judge the passage of time took both waking and sleeping timekeeping into account.

This experiment was conducted in 1931 and 1932. It involved placing two people, one at a time, in an isolation room at Cornell University and keeping them there for extraordinary lengths of time to see how well they could track the passage of that time. The room was simple: a bed, a table, a chair and a washstand. Toilet facilities, water and food were outside in a hallway. Communication between the subject and the experimenters was allowed via a buzzer and a "speaking-tube" whenever the subject wanted to go out into the hall. This warning communication prevented any chance encounter between subject and experimenter that might give time clues to the subject.

The experimenters monitored the subjects' activities and in turn the subjects kept a running estimate of how much time had passed. There were differences between the two. The first subject, Mr. A. D. Glanville, was asked to report what time he thought it was roughly every half-hour to an hour, record his reasons for making that guess and occasionally leave what he had written in the hallway. Glanville was also allowed to write whatever he wanted, resulting in the production of several letters and even part of a story. The second subject, R. B. MacLeod, was kept on a shorter leash: because the experimenters felt that Glanville had spent too much time writing, MacLeod was instructed to do none at all. He dictated his time estimates into a recorder, only when signalled to do so, and was to report the times not as clock times but as estimates of the number of minutes that had passed since his last report.

There were two impressive things about this experiment. One was the total time each volunteer spent in the room: Glanville, eighty-five hours and fifty minutes, and MacLeod, forty-seven hours and fifty-six minutes. That's the equivalent of a weekend for MacLeod and a long weekend plus for Glanville. But their time estimates were truly amazing. Even though both wavered in their estimates during the experiment, sometimes drifting three or four hours or more away from clock time, both timed the end of the experiment with incredible accuracy. Glanville was only forty minutes off, MacLeod twenty-six. Both errors were less than 1 percent! This, even though Glanville admitted in his debrief that he wouldn't have been surprised if he had been off by half a day in either direction.

Many intriguing details emerged from this study, but one that perhaps has the most relevance to the ability to wake up at a specific time is the observation that the subjects' estimates of clock time ("I have reported the time as 8:30 A.M., principally because it felt like 8:30 A.M.") were consistently more accurate than their estimates of how much time had passed since the last guess ("I should judge that it is about two hours since I had lunch . . . I shouldn't be surprised to find myself several hours off in either direction"). As the authors of the study pointed out, the subjects' "general orientation" in time was a lot more accurate than their judgments of the length of specific stretches. Some sort of general time orientation during sleep—a kind of biological clock—could explain waking at the right time more easily than somehow keeping track of an accumulation of elapsed times, say from one waking to the next.

A glimpse of that biological clock was provided in a study published in the journal *Nature* in 1999. A team at the University of Lubeck in Germany reported that subjects who had been warned that they would be wakened at 6:00 A.M. experienced a dramatic rise in the stress hormone adrenocorticotropin starting an hour before waking. Those same subjects, when wakened without warning at the same time, had no such elevated hormone levels. This hormonal change might be the trigger for self-awakening. Adrenocorticotropin was already known to be involved with wakening, because levels of it and another hormone, cortisol, rise slowly in the blood near the end of sleep and reach their daily peak right at the time of waking. They are also secreted in anticipation of stress. In fact in this experiment both hormones rose immediately in those who had been wakened without warning at 6:00 A.M. The authors suggested that this double hormonal increase was a stress response to the surprise awakening.

This experiment provided the first-ever evidence that waking before the alarm goes off actually has a chemical basis, and one that has its origins at least an hour before the target time of waking. In that sense this is the first "hard" evidence that there is a clock in our brains, and that it is running night and day, sleeping or waking. You could even speculate that those times when waking on time is really important are unusually stressful, resulting in elevated levels of these stress hormones anyway, and that, combined with the natural rise in adrenocorticotropin in anticipation of waking, could create a very restless night, full of premature awakenings.

It is a remarkable, if elusive, phenomenon. It demonstrates that the brain can not only remember a command during sleep but that it can—in some as yet undetermined way—link that command to a ticking biological clock and cause waking. How would it work? There are only the vaguest of suggestions, like the idea that at least two systems are at work here, one that creates a lighter sleep in anticipation of having to wake up, and another that checks the times on the biological clock periodically and then connects the two to cause waking. The fact that it all happens during sleep is a puzzle. I'd guess that unlike Winslow Hall, most of us who think we can wake ourselves at an appointed time would be less able to guess the correct time during the day when we're awake.

Maybe that should be the next set of experiments. They would be much less difficult to manage than experiments on the time of waking, could be done with dozens of subjects at a time, and might reveal just how accurate our internal timekeeping can be, even when it has to compete with the ongoing activity of an awake brain.

The Tourist
Illusion

One of the most striking—and unsettling—experiences I've ever had is to travel by car to a place I've never visited before, then travel back along the same route. The trip out always seems to take longer than the trip back—not just a little longer, but much, much longer. I vividly remember travelling by car a few summers ago to a town in Eastern Ontario that I had never been to before, expecting that each twist and turn of the lakeside road I was following would be the one to finally reveal my destination, hoping that the crest of every hill should provide a glimpse of a church steeple, yet I drove and drove and drove with nothing but farmers' fields on one side and docks and beaches on the other. By contrast, the trip back the next day was a very different story, a brisk little drive completely free of tension, insecurity and surprise.

The effect is most striking when you're driving, but it will work if you're riding a bicycle or even walking; speed isn't as crucial as having a destination. It's too awkward to call this the "time there/time back discrepancy," so I've named it the "tourist illusion"—it works when the outgoing trip is a first-time excursion to a place an uncertain distance away and the return trip must retrace your steps. As long as the speed you travel remains more or less the same, the effect is profound, even shocking, and represents a dramatic disruption of the human ability to keep time.

Unfortunately, explaining exactly why the trip back is so much speedier is a little tricky. For one thing, unlike visual illusions, where you can revisit the experience whenever you want, this illusion is different every time you experience it, and by definition, can't be exactly repeated. Second, the psychology lab doesn't provide a lot of help in interpretation, at least partly because the experience usually lasts hours, an uncomfortably long time to run an experiment— even undergraduate psychology students have their limits.

However, there is some research that suggests what might be going on. First, a crucial part of the tourist illusion seems to be ignorance of exactly where and how far the destination is. It doesn't have to be *complete* ignorance—you usually know roughly where you're going and when you're likely to get there—but it doesn't work nearly as well if you have detailed knowledge, like a map or a set of landmarks. The voyage must be somewhat mysterious, which demands that you pay close attention to every feature of the landscape along the way.

From my own experience I'd say that anticipation, expectation and attention all might have played a role in the dilation of time during the outgoing trip, while the absence of all these and the recognition of now familiar features of the drive were likely an influential feature of the trip back. The question is, do these factors really play a role, and if so, how do they influence our perception of time?

We do have internal timers—biological clocks—but they do not run perfectly steadily. What's going on in our minds, the drugs we take and even our general health can affect our judgment of passing time. One of the most vivid demonstrations ever of the alteration of such a clock was provided by biochemist Hudson Hoagland in the 1930s, when he noticed that his wife, bedridden with a fever of 40°C, seemed to be misjudging time dramatically. She asked him to run an errand to the drugstore, and even though he was gone only twenty minutes, she insisted that he must have been away much longer. Hoagland, ever the scientist, then experimented with her sense of time by asking her to judge the passage of a minute by counting to sixty, second by second. The higher her fever, the faster the count. Apparently the fever sped up her biological clock (chemical reactions do accelerate with temperature) and her clock had therefore ticked over more than usual when he went to the drugstore, leading her to believe he had been gone much longer than twenty minutes.

I suppose if you had a fever on the outgoing trip but had recovered by the time of the return trip, you would experience this very illusion. But obviously that isn't the basis for the majority of these experiences, and in fact it's unlikely that this particular timer is even involved in this illusion. Rather than the disruption of a relatively precise biological clock, it seems to involve something a little vaguer, a "sense of time," not an actual timekeeper.

The experiments that come closest to mimicking the tourist illusion are those that have tried to connect memory and attention to our estimates of the passage of time. In the late 1960s, psychologist Robert Ornstein used a computer metaphor to explain how we judge "duration," the amount of time any experience or series of events appeared to take. He argued that if you feed information steadily into a computer, and you know how that information is stored *in* the computer, you can estimate how much disc space (he called it "storage" space) is occupied by that information. The sheer amount of information, and its complexity, will dictate how much storage is needed. Ornstein then argued that as storage size increases, the apparent duration of the process will increase along with it. That would mean that the outgoing trip, being novel, was full of new information, in contrast to the return trip that simply revisited the same landmarks.

Ornstein was careful to emphasize that the way information is stored is as important as the amount. For instance, the number 186719452002 would take some effort to memorize, unless you knew that it was three dates: Confederation, the end of World War II and Canada winning hockey gold in Salt Lake City. If you knew the code, you'd require less storage space; if you had to memorize many such sets of numbers, you'd estimate the time you had taken to do it as shorter than if you had had to store each one as a set of unrelated digits.

One of my favourite Ornstein experiments is a good illustration of what he's getting at. Volunteer students listened to two different audio tapes containing a variety of everyday sounds, like a typewriter key striking the roll, a quick turn of the same roll, zipping a zipper, tearing a piece of paper and blowing across a beer bottle (not just any beer bottle, but the tasty Czech beer, Pilsner-Urquell). Ignoring the fact that students today wouldn't even recognize the first two sounds (but would be completely familiar with the last), Ornstein arranged them in two different ways: on one tape each sound was repeated twenty times in a row until all ten sounds had been heard; on the other tape the sounds were played randomly until they totalled the same two hundred. In both versions the total amount of information was the same; it was just arranged so that one version was simple and the other wasn't. The random series was judged to take about a third longer than the other, confirming the idea that the complexity of the information dictates the apparent duration.

Ornstein's experiments asked participants to judge how long the experiment had seemed *after* they were finished. This is important:

there is a crucial distinction between judging duration after an experience versus judging how long the experience is taking while you're involved in it. It's important to separate the two because at first glance they seem contradictory.

"Time flies when you're having fun." This well-worn saying suggests that the more interesting, absorbing (and enjoyable) the activity is, the faster time appears to pass. This should mean that the apparent duration of that activity would be less, not more, but Ornstein's experiments (and many others since) have shown that when you think back to that activity, it will likely seem longer, not shorter. How can an experience seem short at the time but long in retrospect? Because, as the great American psychologist William James pointed out more than a century ago, the judgment of time as it's happening draws heavily on your attention; the judgment of time past, however, relies on memory, and the two are very different.

Most neuroscientists agree that the brain resources we can devote to attention are limited, and worse, are shared with the resources you need to keep track of time. If most of that capacity is focused on processing complicated information or paying attention to many diverse bits of information, complex or not, you will not be able to pay close attention to timekeeping.

That's why time flies. When your mind is focused on something other than the passage of time, you are fooled into thinking that less time has passed. Of course when you're thinking of nothing other than time, it expands, as captured by "A watched pot never boils." (There actually was a paper published in 1980 in the *Bulletin of the Psychonomic Society* called "The Watched Pot Still Won't Boil: Expectancy as a variable in estimating the passage of time." Students watched a pot boil and then estimated the time that had passed. Their estimates were—not surprisingly—longer than estimates of students who had not watched a pot.) We know from experience that this phenomenon works over long periods of time; children waiting for Christmas, with all their mental resources focused on the slow approach of the big day, are the perfect example—time drags interminably. Yet the hours of Christmas morning flash by.

Experiments have shown that time can easily be confused with seemingly unrelated things like speed and distance travelled. In the 1950s, psychologist John Cohen put passengers in cars with blacked-out windows and told them they were going for a trip and at some point during the trip a bell would ring. After the trip the passenger was asked to estimate the duration of the trip, the speed and the distance covered. The surprising discovery was that all three were connected. If the bell rang at exactly the halfway point *in time,* but one-half of the journey had been speedier than the other, the passenger would judge that speedy half to have taken longer. This experience can be mirrored in the lab if a person faces a set of three lights spaced so that the first two are relatively close but the third is ten times farther away. If the three lights are flashed in sequence at exactly equal intervals, onlookers judge the time between the second and third light (the two farthest apart) to be significantly longer.

But again these are examples of judgments made of time as it is unfolding, rather than time past. The situation is very different when you're looking back, partly because when you're asked to keep track of time as it's passing, you're focused on time and little else. When you're asked to estimate time that has passed, it's less about attention and more about memory. The more events that you remember having happened during a specific period of time (as Robert Ornstein showed), the longer that time will seem to have been.

Richard Block at the University of Montana is an acknowledged expert in this area, and he has added one important idea. He contends that it's not just the number or complexity of events you remember that influences your perception of how much time has passed, but the context. In this case context can be the way the information is being processed, the emotional state of the person or even the physical setting.

For instance, Block has run many experiments over the years that have convinced him that if you perform two tasks that are pretty much identical, you will, at some later point, judge the first of the two to have taken longer. Even though this conclusion has been drawn from psychological tests conducted in settings very different from the tourist illusion, it sounds as if it might be a significant part of the explanation. Block has pointed out that it's a little surprising that the first of two experiences should in retrospect seem to have lasted longer, because as memory fades, events should drop out of the earlier version first, making it seem shorter. But because the first introduces a new context and the second simply continues it, the first experience seems longer.

To underline this point, if the context is changed but the amount of information stays exactly the same, estimates of time spent change accordingly. One such study asked students to do two tasks, one of which was simple—identifying which of a series of words are capitalized—the other hard (or at least *harder*)—picking out those words that described a part of the human body. Those students who did just one task, either the simple or the hard, judged less time to have passed than those who had to switch back and forth. (If our sense of time passed depended strictly on the amount of information that we processed, those students who had performed only the hard task should have estimated more time to have gone by than those who switched. But they didn't—another blow to the idea that our estimate of duration depends on the amount of information that we remember.) Block has found the same effect when he temporarily moves students out of the room and into the hall between tests, or into a different room altogether. The more changes, the longer the estimated time. By changing the surroundings or the way information is presented, he has been able to eliminate the illusion that the first of two experiences always seems the longer.

With all this as background, it's much easier to analyze the tourist illusion. Whether driver or passenger, it's easy to imagine the situation: you know you're heading toward a goal, but you have no idea exactly when you will reach it or, indeed, what it looks like—it's a name of a place and not much more. As the experience unfolds, you're devoting most of your attentional resources to the changing landscape, your own fatigue and anticipation, each new street sign, the impatience of the children in the back seat, your explanations to your partner as to why it's taking so long, your questioning of your partner as to why it's taking so long . . .

There are innumerable sights, sounds and feelings to be recalled. Each of these separate memories comes with a context automatically attached. Richard Block's studies have established that there's no need to consciously commit those contexts to memory—they just come along for the ride. In fact it's probably hard to separate memories from context. Every change—a glimpse of something new around the next bend in the road, your sudden realization that your dinner reservation is creeping ever closer—is an example of both. Even the fact that you see some things, hear others and talk about others—all add to the growing pile of contexts. All of these will affect your memory of the outgoing trip but won't actually be influencing your feeling of how much time is passing as you're travelling. While it sounds like all the attention to ongoing events should make time go by unnoticed, the fact is that that attention is all *about* time: "When do we get there?"

The return trip is an entirely different story. You've been there before, like the students sitting in the same classroom for the second round of psychological tests, and that alone will shorten your estimation of the amount of time that has passed. It will be much more difficult for each new scene or event to qualify as a new memory:

the old oak tree on the hill is already stored as a memory of the outgoing trip, and although you will recognize it, it won't qualify as a new memory. You don't even bother reading the same road signs that preoccupied you on the way out. Attention is focused on other things, now that the crucial part of the trip is over.

Of course you could easily eliminate the illusion by taking a different route back, but there's a trade-off: you would be adding a new experience to your memory, but you would be losing the opportunity to experience a compelling illusion. There are always new road trips, but great illusions come by only once in a while.

The ATM
and
Your Brain

I hope I'm not the only one who has experienced the following confusion: I want to take money out of an ATM, so I have to insert my bank card into the slot to unlock the door. But it seems that ATMs are not only different from gasoline pumps, parking ticket dispensers and other objects that accept your card, they're different from each other. Some of the slots are horizontal; some vertical. Some demand that the black stripe should be uppermost on the left-hand side, others that it must be horizontal, downward-facing and on the right. If you've experienced the sometimes lengthy pause I have deciding how to position the card correctly, relax: thirty years of research attests to the fact that your brain must perform some clever mental gymnastics to get it right.

There is more than one way to solve this problem. One is simply to try every single orientation of the card until one works, but that's tiresome, potentially embarrassing and a cop-out. What do you have a brain for if not to figure this problem out and then get it right the first time? To do that you have to accomplish what is called "mental rotation," shifting the position of the card around *in your mind,* before sticking it into the slot. This process has been investigated in detail, and the amazing thing about it is that it turns out to be very lifelike: turning the card around mentally is almost exactly like turning it around in your hand.

The first experiments that demonstrated exactly what's going on were done in the early 1970s. Psychologist Roger Sheppard and his graduate student Jacqueline Metzler showed pictures of pairs of funny little 3-D objects to people and asked them whether the two objects were really the same one in different positions or were two different mirror-image objects. The task was something like being shown two gloves at different angles and being asked to judge whether they are a pair or two gloves of the same handedness. However, the objects in the Sheppard-Metzler experiment were more difficult than gloves: gloves are two-dimensional, but these objects were 3-D, chains of ten little cubes, strung together in odd ways. One might look like the letter "L" but with another arm at the top sticking out of the page. If you have ever taken any kind of intelligence test you've likely encountered such objects. They can be fiendishly difficult to deal with, but they sure are popular among psychologists who investigate mental rotation—they're still the experimental object of choice today, more than thirty years after Sheppard and Metzler's experiments.

The task in these experiments was to take one object and imagine it rotating until it more or less lined up with the other. At that point it's possible to judge whether the two are actually the same object or different. I'm not sure anyone expected the results of these original experiments: the further an object had to be rotated in a person's mind, the longer it took to make a decision. In other words, rotating an object in your mind takes time, just as it does in the real world.

Apparently nobody can take one of these 3-D figures and simply snap it into the right orientation; it has to be turned just as if you were holding it in your hand. It's worth remembering that in our minds there's a long list of things we can do that have little or no resemblance to the way things work in the real world, including dreams, transporting yourself back to the place you grew up or imagining what your vacation is going to be like. But in this case your brain seems not to be able to leap to a conclusion, but instead painstakingly moves the object—or, rather, the picture of the object—just as it would if it were there in front of you. Sheppard's subjects claimed they imagined turning one of the objects until its top arm paralleled the top arm of the other, then they'd check to see if the arms at the other end matched or not.

Sheppard pointed out that it's not just human beings that can do this—he told a story of a German shepherd that charged through the space left by a missing board in a fence to retrieve a long stick. Having grabbed the stick in his mouth, he started back through the narrow space only to stop, pause and turn his head so that the stick would go through the narrow gap. Not only could the dog visualize what would happen if he didn't rotate the stick but presumably he did it without talking to himself, so language isn't necessary for performing this trick. (That's a little surprising to me, considering that I find myself talking at the door of the ATM quite often.)

The consistency of the timing was consistent. The participants took anywhere from about a second to make their decision (if the objects were in the same orientation) to nearly four and a half seconds (if they were the maximum 180 degrees apart). By simple arithmetic this gives a rotation speed of just over 50 degrees per second. Sheppard pointed out that this regularity of timing excluded other possibilities for what his subjects were doing. For instance, they couldn't be coding the objects as "two blocks, right bend, four blocks, upward bend, two blocks, left bend, two blocks" and then comparing the codes of the two objects, because it would take the same amount of time to judge each pair no matter how different their orientations were.

When I stand at the ATM door I'm comparing the orientation of my card with that of the slot, and experience tells me that it's a two-step process: first, hold the card so that it lines up with the slot, then rotate it or flip it side-to-side, or both, to match the little diagram and get the door open. These experiments thirty years ago partly explain what's going on in my brain to do that: I'm performing a mental rotation of the card to line it up. But there's more to it than that.

Recent experiments have identified the parts of the brain involved in mental rotation. One is the parietal lobe, a large section of the brain at the top of the head. (If you wear a baseball cap pushed back off your forehead it nicely covers the parietal lobes.) This part of the brain is involved in judgments of your position and orientation in space. Damage to the right parietal lobe is notorious for causing a perceptual dysfunction called unilateral neglect, the lack of awareness of anything on one's left side. Some people with such neglect have to teach themselves to eat all the food from the right side of the plate (the only food they're aware of) and then rotate the plate so that what originally escaped their attention is now apparent. (One woman, rather than turning the plate, learned to turn her chair so that she could approach the plate from the left.) Neglect is a serious disability caused by damage to the brain, but it illustrates just how important for the perception of space and the objects in it the parietal lobes are.

Neglect of the left side, caused by damage to the right parietal lobe, is much more common than neglect of the right, suggesting that the right hemisphere dominates the perception of space. Damage to the left parietal lobe seems rarely to cause neglect of the right side, likely because the all-powerful right hemisphere is compensating for it.

The consensus in the 1990s would likely have been that the right hemisphere is in charge of mental rotation too, but several studies have suggested that this is too stark a picture. One research group has claimed that the two hemispheres split the job: in their experiments the right seemed best at maintaining an internal image of the original object while the left rotated the second object until they fit, or didn't. Daniel Voyer of St. Francis Xavier University found evidence that the right hemisphere is dominant when the brain first tackles the task, but that with training the routine becomes localized in the left hemisphere. In that case, the right hemisphere would be the one that deals with unfamiliar routines, the left with routines that have been worked out and streamlined.

It's a complex picture. One man suffered a stroke on the right side of his brain that affected his ability to mentally rotate his hands but left his ability to mentally rotate other objects untouched. So he would perform reasonably well in matching the Sheppard objects, but had trouble when asked if an image of a hand with two fingers sticking out matched another image with two different fingers extended. This one case suggests that mentally rotating parts of the body likely engages a different part of the brain than rotating objects, which is not really surprising: our bodies and the space in which they move are fundamentally different, and it would make sense that mental images of our arms, legs, feet and hands (and control of the way they move) should be the responsibility of a unique brain area.

I'm likely using all of these dedicated brain areas as I stand before the ATM door because I'm mentally rotating an object—my bank card—to make the card's orientation match the diagram beside the slot, but at the same time I'm checking the relationship between the card and my hand, then moving my hand into the right position to insert the card. Moving my hand is a new step in all this—it has only been *mental* rotation so far—but the most startling new development in the understanding of mental rotation over the last few years has been the discovery that the parts of the brain that control movement, the so-called motor areas of the brain, actually do play an important role in mental rotation, even if that rotation doesn't require you to lift a finger.

This sounds odd. After all, imagining the rotation of some object on a computer screen doesn't imply in any way that you might be moving that object, yet that's exactly what your brain seems to be planning. In one of the most elaborate experiments that demonstrate this, Andreas Wohlschlager of the Max Planck Institute had subjects sit in front of a computer monitor that showed them images of the typical Sheppard objects, but in this case they were confronted with two tasks: one was the usual, to decide whether two of those objects were the same or different—the familiar mental rotation test. Subjects identified whether two objects matched or not (after rotating them) by pushing a button with their left hand. But they then had to use their right hand to turn a video-game-like trackball in the direction indicated by an arrow on the screen. The trick here was the order of events: first an arrow popped up indicating which way to turn the trackball, but the subjects weren't to do anything about it right away; they had to store that direction in their memory while they switched over to the mental rotation task, using their left hand to push the button to indicate whether the two objects matched, then finally switch back to their right hand to move the trackball in the direction the now long-gone arrow had indicated.

Wohlschlager used this somewhat complicated process to answer a straightforward question: if you are already planning to make a movement (as the people in his experiment were when they saw the arrow) but you haven't actually done it yet, does that interfere with the ability to do mental rotation? In other words, if your brain is thinking about making a real movement, does that make a second, imagined movement trickier? The answer was yes. People in this experiment took longer to do the mental rotation necessary to decide whether objects matched if they were already planning to make a hand movement that required rotation as well. Especially difficult was the case where their planned hand movement was counter-clockwise, but the mental rotation was clockwise. The two clashed and the reaction time was even slower, providing convincing evidence that the motor areas of the brain are involved in some sort of covert planning of movement even if you're just rotating objects in your mind.

Wohlschlager's experiment is just one of many to show this. Some studies using transcranial magnetic stimulation, a technique that uses magnetic fields to temporarily inactivate part of the brain, have shown that inactivating motor areas in the left hemisphere slows down people's abilities to match rotated shapes. The curious thing is that the motor area only starts to get involved long after (at least in thousandths of a second) those areas responsible for analyzing the shape of the objects. Shape first, planned movement second.

Research published just as I was writing this chapter added one curious wrinkle to this story: our brains apparently pay more attention to objects that can be grasped than to anything else. Researchers at Dartmouth College showed that if people were shown two objects simultaneously, one that we're used to holding, like a fork or a pen, the other ungraspable, like a cloud, more brain activity is directed toward the graspable object. Even odder is the fact that it is the right side of the brain that becomes excited. The team is still checking out whether being right- or left-handed matters. The question is, has the near-ubiquitous bank card attained the status of a "graspable" object yet, or does the fact that you can hold it practically any which way (part of the problem I'm trying to solve here) prevent it from exciting those parts of the brain that recognize the handle of a knife or the pen in the upright position? If it is "graspable," then the neuronal action will be even hotter.

Now back to the ATM to sort all this out. I have my card in my hand, intent on moving it into the right orientation to plug it into the slot and get to the machine and get my money. There is chaos in my brain! (Had I been born a female there would still be chaos, but of a different kind. There's good evidence that men and women who are equally good at solving mental rotation tasks nonetheless rely on different brain areas in different sequences. There are many precedents for gender differences like that—even language seems to be organized differently in men's and women's brains.) My parietal lobes are struggling to figure out how to match the card in my hand to the image of the card beside the door. They, and other parts of my brain dedicated to moving my own body parts, are then busting their synapses to match image to card to movement. Once all systems are convinced that the card is the right way round, then it's just the simple matter of shoving it into the slot, an action that likely occupies a mere several million more neurons.

There is one final piece of good news: mental rotation tasks get easier with practice, meaning that as long as the banks are consistent in the placement of card slots for their ATMs, I'm going to be able to do this faster and faster as time goes on. Eventually it will be as if my card is merely an extension of my arm, my right hemisphere will abandon the task to my left, and finally the whole thing will disappear from my conscious mind completely, leaving my brain free for other, more intriguing thoughts. Maybe I'll even write poems when I visit the ATM.

SPEEDING
TO A STOP

A loonie doesn't go far these days, but you can still spin one on a table. Being a largish coin it will spin well enough, although it still doesn't go very far—it just stays in one place, spinning away. It wobbles as it spins, like a top slowing down, and as its wobble becomes more and more eccentric and the coin leans over farther and farther, the sound, which had been thrumming along with the spin, becomes a higher-pitched whirr as the coin accelerates, then stops abruptly with a loud "vhrummpp."

Many academics have thought about this—although because they are scientists they prefer to consider idealized discs rather than actual coins. They wonder why the disc behaves the way it does, especially in the final few moments, which they like to refer to as a "shudder." To understand why the experts should be so fascinated with something that you can demonstrate with nothing more than a coin and a tabletop, you have to distinguish between two different effects.

They are *spin* and *wobble*. If you take a loonie between thumb and forefinger, stand it on its edge, then spin it, it will rotate on its edge, sometimes for quite a while. In a way that's not surprising—after all, you can stand a coin on its edge easily. A British mathematician once calculated that there is a finite chance—small but finite—that you could toss a coin into the air and have it *land* on its edge. That is spin. However, the coin inevitably starts to wobble. It has lost its initial stability, it is no longer on its edge but rather on the *edge* of its edge, and that is the beginning of wobble, a completely different phenomenon from spin.

Wobble is easier to see with a top, which usually has an easily identified spinning axis. No matter what the shape or design of a top, the axis is the part you twist to give it its spin. When it starts to wobble you can easily see that while the top might still be spinning pretty fast, the wobble is tracing out lazy, much slower but ever-widening circles in the opposite direction. The technical term for this is "precession," and even the axis of the spinning Earth precesses. One of the many results of this is that the current North Star, Polaris, was in a completely different location in the sky thousands of years ago, and wouldn't have qualified as the North Star. But the Earth's axis has wobbled—precessed—since, and so today we see Polaris in a different position. When a top starts to precess, it can't go back: the wobble describes wider and wider circles in the air, and eventually the top tips over, hits the tabletop, and that's the end.

The coin does pretty much the same thing, but its movement is much harder to follow. The spin is easy enough to see, but to catch the wobble in action you have to follow not the axis (it's not obvious like that of a top) but the precise point of the edge of the coin that is in contact with the tabletop. It's this point of contact that wobbles, moving again in those ever-widening circles.

One good way to check this out is to look straight down on the coin as it begins to wobble. The loon (or the Queen) is turning fairly slowly—that is the spin. But if you get down on your knees and look across the tabletop at the coin, you'll see that a single point on the coin's edge is moving around and around on the tabletop. It's moving much faster than the spin in the same sense as the people in a crowd doing the wave aren't going anywhere, but the wave is travelling around the stadium.

However, there is something weird going on here, and it has something to do with the sound I referred to at the beginning of the chapter. It is intriguing because the sound rises sharply in pitch just before the coin topples over. It might even be the first thing that attracts your attention, but if you then take the time to watch as you listen, you'll see that the sound accompanies a steady acceleration of the wobble, which in the last micro-moments before the coin shudders to a halt becomes nothing more than a blur. Does it look like nothing can hold it back, that it would have just kept speeding up if the coin hadn't toppled over? In fact, there might not *be* a limit to the speeding wobble.

In 2000, a British mathematician named Keith Moffatt published a short report in the journal *Nature* about the forces driving the spinning coin, and concluded that the equations that describe that motion predict that as the coin tilts over, spinning slower and slower, the wobble rises to infinite values. Besides the obvious fact that this does not actually happen, it's a disturbing finding to physicists because it signals the presence of what scientists call a "singularity," a situation in which the laws of physics fail.

Very grand things contain singularities, like the Big Bang, when the universe was infinitely small, or black holes, where the force of gravity is infinitely large. But a wobbling coin? Moffatt was intrigued that the equations lead to the conclusion that as the angle the coin makes with the surface flattens (the coin is tipping over) and as it wobbles faster, there comes a time when the angle approaches zero and the rate of rotation approaches infinity.

Because infinity never is the actual result in this case, Moffatt searched for an explanation.

Of all the forces that might stand in the way of infinite wobbling, Moffatt nominated the drag on the coin exerted by the cushion of air between the tilting coin and the surface on which it is wobbling. As the coin gets close to the end of its run, it has tilted so far that it is almost parallel to the surface, and it is easy to imagine that the air between it and the surface is being squeezed against the surface of the coin. Moffatt calculated that the forces the coin then encountered were enough to halt the wobble and prevent the singularity from developing. Best of all, he had an opportunity to play with a spinning disc, all in the name of science.

The disc he chose to test his equations is a toy made by Tangent Toys called Euler's Disk, named after Leonard Euler, the man who invented much of the math used in studies like this. Euler's Disk is like a spinning coin, but it has special properties. For one thing it's big and heavy—more than 400 grams; it reminds you of a stainless steel hockey puck. For another its edges, unlike those of a loonie, have been finely machined and are beautifully smooth. When you buy the disk it comes complete with a mirror to spin it on, and it is capable of spinning and wobbling for upwards of a minute before it stops—a loon is good for ten seconds or so at most. Moffatt's equations had predicted that the Euler's Disk would last a hundred seconds or so, which is well within its range.

However, Moffatt's analysis, rather than sewing up the story of spinning coins and Euler's Disks, created controversy. A number of math and physics whizzes immediately turned their attention to the problem and came up with some radically different ideas. In one case a group of scientists in the United States pointed out that objects like wedding bands exhibit the same accelerating wobble and shuddering halt, even though they can't possibly be squeezing a cushion of air underneath them. They then dealt what you'd think would be a killing blow to the air friction idea by spinning coins in a bell jar after they had evacuated the air from it. The coins behaved in exactly the same way as they had when there had been air in the bell jar, so clearly the absence of air hadn't had the significant effect on the coins' wobble that Moffatt might have predicted.

These scientists suggested that Moffatt's idealized picture of the situation was too simple. It might be true that friction between the edge of a coin or Euler's Disk would be minimal if the point of contact traced a circle on the tabletop, but a number of forces collaborate to make that unlikely, and some of them cause the edge of the coin to be rubbed along the tabletop "in a jerking motion." They went on to argue that this friction would cause the coin to bounce up and down, eventually disrupting its motion so much that it would come to rest. In their view it's all about "edges rubbing against the tabletop." They also pointed out that evidence for the importance of friction in the slowing of spinning discs could be seen with Euler's Disk: when spun on a table rather than on its own extremely smooth platform, it spun for only a few seconds, not the hundred-plus that it's famous for.

Moffatt replied by saying that he realized that there might be other factors involved, that his equations had matched what was seen for Euler's Disk and that equations describing more exotic influences would be difficult or even impossible to solve. Some commentators supported at least part of what Moffatt had claimed. A group at the University of Massachusetts at Amherst took Moffatt's equations one step further by recording high-speed videos of a variety of heavy steel rings and discs as they spun and wobbled. They found that Moffatt's equations did pretty well at describing the way these objects behaved, although their videos prompted them to tune the numbers a bit, but they disagreed over what was finally slowing the coin, eventually causing it to topple over. They argued, as had the critics who spun things in a vacuum, that friction between the edge of the coin and the surface was what did it.

When they spun discs on different surfaces (glass, steel and slate, in order of increasing friction) the equations changed for each, something that shouldn't have happened if air was the key factor. After all, the air was the same in each case. They also did the ring thing, showing that a disc and a ring behave the same, even though the ring traps very little air under it.

It wasn't that these scientists dismissed the importance of the air completely. In fact they calculated that at extreme rates of spin, approaching the singularity, air drag would overtake rolling friction and become the prime factor slowing the coin. But long before that, at least in real life, friction has slowed the coin down to the point of its falling over. They did hypothesize that extremely thin discs might be different, but I got the feeling from their caution about this ("it will be a challenging task to ensure the rigidity and circularity of the disk") that they were thinking about really, really thin discs, the kind only physicists can dream up.

The questions raised about Moffatt's paper and his responses are a good illustration of the sometimes frustrating gap between the precision of physics and the sloppiness of the real world. Most of us just want to know why a wobbling coin appears to be speeding up, but to answer that question correctly requires theories that generate equations that can be tested by experiments, and, as Moffatt pointed out, some of those equations are right on the edge of what is actually known about rigid discs rolling around on smooth surfaces. Equations are beautiful ways of describing what happens, but each set describes only a narrow range of possibilities: this disc, rolling at this speed, on this surface, will behave this way. That description may not hold for the nickel on the kitchen counter.

Even the Euler's Disk is vastly different from a spinning coin. For one thing, the edge on which it spins is not squared off like the edge of a coin but machined to be smooth and curved. That might be stabilizing the disc against the inevitable moment when it loses its grip and falls flat.

It's not all critiques: some scientists have been stimulated by Moffatt to publish their own thoughts about Euler's Disk. Oliver O'Reilly in the Department of Mechanical Engineering at the University of California Berkeley is one of those. He and colleague Patrick Kessler think that the coin's vibration as it accelerates, and the accompanying vibration of the surface on which it is wobbling, play a role both in the rising pitch of the sound it makes and the final shuddering stop. They argue that both the wobbling disc and the surface vibrate, and some of those vibrations will be amplified by the disc. In the same way, rubbing the tip of your finger along the wetted edge of a wine glass will cause it to vibrate, but only certain frequencies of sound will be amplified by the glass. Kessler and O'Reilly suggest that those same vibrations can, as the disc nears the horizontal, actually cause it to lose contact with the surface. The disc is momentarily airborne, but when it lands, it stops.

O'Reilly and Kessler point out that they could go further in their analysis, especially because the simulation they were using assumed that the disc and the surface were rigid, and of course if they were perfectly so they couldn't vibrate. In order to test their ideas about vibration and capture the movement of the disc in ever more realistic detail, they would be obliged to create a new simulation, one that they say would be "very expensive." Add to that Moffatt's comment that determining the frictional forces between disc and surface at the point of contact between them would be a "difficult problem, which, so far as I'm aware, still awaits definitive analysis" and the picture becomes clear. The spinning coin problem is pushing the envelope not only of what's known but of what *can* be known. And that's more than a loonie usually buys.

It's a
Small World
After All

W e've all had the amazing experience that it is indeed a small world: you stop for french fries at a chip wagon in Brussels and the person in line ahead of you turns out to be sharing an apartment with someone who went to the same high school in Edmonton as you did. This is, of course, also called "six degrees of separation," the idea that you can be connected to anyone in the world via a chain of no more than six people, each of whom knows the next person in the chain.

The idea has spawned a play, a movie and games that connect actors (*The Six Degrees of Kevin Bacon*), baseball players, people who might know Monica Lewinsky, and my favourite, a website at the University of Virginia called "The Oracle of Bacon at Virginia," where even the briefest of visits will convince you that Elvis Presley made movies with just about anyone who was alive.

It is not widely known that the first article written about the phenomenon was published in the inaugural issue of *Psychology*

Today, in May 1967. *Psychology Today* was a very different magazine in those days, much more oriented to experimental psychology than self-help. It's unusual to find an original and highly influential article like this in a popular magazine. The article was written by psychologist Stanley Milgram, who had performed a fascinating experiment on what he called the "small world" effect.

Milgram was given $680 (U.S.) by the Laboratory of Social Relations at Harvard University to design an experiment that would test the idea that any two Americans could be connected by a short chain of acquaintances. He chose the cities Wichita, Kansas, and Omaha, Nebraska, as the starting points of the chain of relationships. (Milgram admits in the article that from the point of view of Cambridge, Massachusetts, these cities seemed "vaguely out there on the great plains or somewhere.") Volunteers were chosen in those cities, and their job was to connect to a single individual: for the Wichita study, the target was the wife of a divinity school student in Cambridge, and for the Omaha version, a stockbroker in Boston.

The volunteers were given a document identifying the target person, but were cautioned not to try to contact the target directly (unless they already knew them) but to mail the document to someone they knew on a first-name basis who would be "more likely . . . to know the target person." Everyone who handled the document added his or her name to a list that became the record of the chain of contacts.

Although Milgram and his colleagues had no idea whether the experiment was even going to work, evidence that it would came quickly: four days after the experiment began, an instructor at the Episcopal Theological Seminary in Cambridge approached the wife of the divinity school student and said, "Alice, this is for you." At first the experimenters thought this envelope had mistakenly never left Cambridge, but the list of participants in the envelope confirmed the small-world effect: the chain of contacts had been a wheat farmer, a local minister, the teacher in Cambridge and the target. In fact, there had been only two degrees of separation between the farmer and the target.

This chain turned out to be one of the shorter ones. They varied from two to ten intermediaries, with five being the median and six the most frequent. A more typical example was the chain that included a self-employed person in Council Bluffs, Iowa, a publisher in Belmont, Massachusetts, and six more people, all of whom lived in Sharon, Massachusetts, home of the target stockbroker. Why did this chain bunch around the target? It is just one example of the many puzzling results that came out of this initial, relatively crude experiment.

Milgram himself is a story worth telling. He is much better known for his experiments showing that people could be easily persuaded to press a lever that they thought was delivering a powerful electric shock to a volunteer, just because an experimenter was ordering them to. The experiment was a set-up—no real shocks were delivered and the "volunteers" apparently moaning in pain were actors—but the experiment itself was shocking. I have watched videotapes of it; the most startling thing is to see the real anguish suffered by subjects as they plead with the investigator (one of Milgram's colleagues) to allow them to stop administering the shocks, yet turn right around and *increase* the voltage when told that they must.

Yet Milgram should also be remembered—more than it seems he is—for this first small-world experiment, a study that gave rise to the current efforts to better understand human communication and social networks in all their variety, from the internet to the transmission of disease. But as intriguing as this experiment was, did the results justify the conclusion that there are no more than six degrees of separation among all of us? A psychologist named Judith Kleinfeld at the University of Alaska reviewed Milgram's files on the experiment from his collected papers at the Yale University library, and she thinks that *any* conclusion from the experiment is premature. Kleinfeld argues that the reason the idea of six degrees of separation has been so wholeheartedly embraced is that people want to believe—in a sort of 1960s "all you need is love" way—that we are connected to each other (she calls it a "heart-warming parable"), and she wonders what the significance of such connectedness would be, even if it were true.

Kleinfeld's initial excitement at finding the original material from Milgram's experiment soon turned to disappointment. What she found was, in her words, "disconcerting." For example, remember the first experiment, the one requiring the delivery of a letter from Wichita to the wife of the divinity student in Cambridge that yielded the dramatic moment that Milgram highlighted in his *Psychology Today* article ("Alice, this is for you")? That letter arrived having passed through only two intermediate links, but what Milgram doesn't acknowledge in his article is that this was one of the only successes. In that experiment, sixty people had been recruited to participate, fifty had begun the chain by handing the letter to someone else, but only three letters had actually been delivered to the recipient in Cambridge. This information is contained in an undated paper in the Milgram archives. The average number of people involved in the three successful chains? *Eight.*

The second experiment, the one that began in Nebraska and ended with the Massachusetts stockbroker, did better. Milgram presented a graph of these results in his *Psychology Today* article, showing that the number of intermediaries in this case was most often six, hence the six degrees of separation. Even so, only 28 percent of the chains were completed (44 of 160). In addition, it appears that the people chosen to start the chains were anything but a random sample. In the Nebraska experiment, most of the participants were either blue-chip stock owners or people chosen from a mailing list. Obviously the stock owners were better placed to be able to connect—eventually—with a stockbroker in Massachusetts. In addition, people whose names are on a mailing list are more likely to be higher income and well connected, again making them better candidates for a successful experiment. Yet even with these apparent advantages, the Milgram experiment achieved nothing better than about a 30 percent success rate, a strikingly low number for a study that has given rise to a widely held belief about the connectedness of human society.

It sometimes happens that an original thought-provoking experiment, even if it is less than perfect, gives rise to repetitions and replications that, by supporting the original findings, paper over its flaws. Not this time though: Judith Kleinfeld could find precious little evidence that anybody actually replicated the Milgram study in all its details. This isn't so unusual. Most researchers understandably want to put their own stamp on a study, and literal, step-by-step replications (something that we're always taught in school is part of the scientific method) are rarely done. But in this case the lack of replications—and results—was stark. In one such effort, in 1968, no statistically significant conclusions could be reached because the number of respondents was less than 20 percent; in another, only 7 percent of letters made it from sender to receiver, although, as Kleinfeld points out, this study was hampered by the researchers' having so little money at their disposal that they couldn't provide stamps or envelopes to the participants! A study published in the summer of 2003 updated the technology to email, but still resulted in only about 1.5 percent completion of chains.

Judith Kleinfeld came to the conclusion that Milgram wasn't trying to mislead but was more likely to have become "beguiled" with the idea that it's a small world and that we truly are connected to anyone else on the planet through a short number of intermediaries. She also makes the point that the rest of us might be similarly inclined. She admits to being amazed at how entranced people are with their own small-world experiences and how strongly even skeptical academics believe that they do indeed live in a small world, even though there really isn't much evidence to support the idea. Kleinfeld has come to believe that there are both personal security and religious reasons for wanting to believe in a small world.

So it may be that six degrees of separation is a myth. It's also possible that the ultimate number, if it could ever be calculated, might indeed fall very close to six—we just don't know. (In fact there is an argument that if people know as many others as some studies have suggested—see the next chapter—then *four* degrees of separation is more likely.) But that uncertainty hasn't impeded research. The idea of six degrees of separation is alive and well and flourishing in social science departments around the world, and some recent approaches have taken a much more abstract approach than Milgram's original study.

One of the most celebrated advances in small-world research was published in 1998 by Steven Strogatz and Duncan Watts. They took a mathematical approach to the subject by comparing two kinds of networks (these would be social networks when applied to people, but as Watts and Strogatz showed, they could just as well describe simple nervous systems or electric power grids). They drew a distinction between two kinds of network: one, an "ordered" network, is tightly and locally organized, with each person (or nerve cell or power plant) having exactly the same number of connections, all tightly clustered around each other; the other, a "random" network, is, as the name suggests, totally arbitrary, with connections all over the place. Most personal networks, the kind that you could draw by plotting your friends on a piece of paper and then connecting them all to you and to each other, are highly ordered—but not completely. Most of your friends likely know each other as well as knowing you, so the lines of acquaintanceship cross and recross and the network takes on a "cliquish" character. But if one of your friends moves to another city and establishes a new circle of friends there, the tight organization of the friendship network has been broken. Watts and Strogatz simulated this sort of event with idealized, computerized networks. They began with complete order and gradually introduced randomness by rewiring the network, one small step at a time. Rewiring amounted to taking a connection or two, detaching them from their nearest neighbour, and reattaching them to a distant point in the same network. They were stunned to find that it wasn't long before they had created a network that had the attributes of a small world. When Watts and Strogatz applied their method to real networks, including the power grid and nervous system mentioned above and a list of actors in feature films, all three seemed to be well described by this hybrid kind of network they had created.

Rewiring only a few of the links between members of an ordered network did two things. First, it created the possibility of instantaneous long-range communication within the network. As long as any network is highly ordered and clustered, it takes a long time to make a connection between, say, Toronto and Medicine Hat. But introduce a few wild card connections (including one that might leapfrog the country to land in southern Alberta) and that long-range communication suddenly opens up. At the same time, however, most of the network remains clustered, preserving the essential structure of typical personal networks. The beauty of this approach is that it goes some distance to explaining why the number of degrees of separation between us all might be as few as six. We have our core group of friends, but we also have those long-distance ("random") connections that bring us much closer to any target person in the world.

The easiest networks to study are those that are created through collaboration, like scientists who co-author reports in scientific journals or actors who have appeared together in movies. The advantage of such networks is that the collaborations are on record, in these cases either in the scientific literature or at the Internet Movie Database (www.imdb.com). Admittedly the nature of the relationships varies in these networks: scientific collaborators have, for the most part, met or even worked together in the lab; actors might simply have been thrown together on the whim of the director. Nonetheless, they provide an unusually detailed record of people coming together with other people.

In perhaps the most unusual application of this research, Spanish scientists have analyzed Spider-man's world—the networks of relationships among superheroes in Marvel Comics—and published their results in the February 11, 2002, issue of a journal called *Preprint.* Their reason for examining what they call the "Marvel Universe" was an academic one: they wanted to know if the mathematical descriptions of real collaboration networks would also apply to this totally fictitious one. In other words, do real human acquaintanceships have some sort of mysterious but quintessentially human character, or can any network exhibit the same sort of mathematical behaviour? They have found that superheroes are connected to each other in similar ways to the rest of us ordinary humans, with one curious but significant exception.

The Marvel Universe is truly Marvel-ous. For one thing, while many of the superheroes who inhabit it are household names, like Spider-Man, the X-Men, Captain America and the Fantastic Four, there are many, many others who, though less well known, have complex ties to countless other superheroes. The Spanish researchers cite the example of "Quicksilver, who appeared first as a member of Magneto's Brotherhood of Evil Mutants in the early issues of Uncanny X-Men, then he became a member of the Avengers and later of X-Factor, to end as leader of the Knights of Wundagore; he is also the son of Magneto, the twin brother of the

Scarlet Witch, and he married Crystal, a former fiancée of the Fantastic Four's Human Torch and a member of the Inhumans."

The rule of thumb here was that any two superheroes are considered connected if they appear in the same comic. Although the fact that the superheroes have been created by Marvel's Stan Lee and his writers and so might appear at first glance not to represent a "natural" collaborative network, in fact most who appear in the same comic have had strong ties to each other, whether cooperative or antagonistic. In that sense they are more like real-life scientific collaborators than actors. The database the researchers used was Russ Chappell's "Marvel Chronology Project" (www.chronologyproject.com), an effort to record every single appearance by every Marvel character in chronological order: so far, 96,000 appearances by more than 6500 characters in 13,000 comics.

The sheer volume of the Marvel Universe isn't as interesting as its details, such as the fact that the average character appears in about fifteen comics, that the record for the greatest number of appearances is held by Spider-man, at 1625, but that the superhero with the greatest number of collaborators is Captain America, at 1933.

The Marvel Universe turns out to be more than just a random set of connections among characters: this is as it should be. Any true collaboration is more tightly knit than random because characters interact with the same people over and over again, rather than collaborating on the basis of the flip of a coin. If the Marvel Universe were random, the number of connections each character would have with others would be 175; the actual number, by contrast, is 52: superheroes collaborate more often with the same people than either actors or scientists do.

The greatest distance in the network between two characters is five; in other words you can connect any two superheroes by a chain of not more than four intermediaries, a close fit to Milgram's six degrees of separation. That is the greatest distance; any two characters *on average* are only two partners away from each other. They are a close-knit bunch indeed. And while Spider-man may capture the imagination these days, Captain America stands at the centre of the superhero network: on average he is only 1.7 characters away from any other.

But there is one curious oddity about the Marvel Universe: it doesn't "cluster" in the way that real-life networks do. In most networks, two people who share a common friend are also likely to know each other: this is called "clustering." For instance, in the movie world two actors who have collaborated with a third—even in different films—are more likely to have appeared in the same film than two randomly chosen actors. Real networks work that way: they branch out from existing relationships and exhibit clustering that can be measured mathematically. By contrast a random network would exhibit very little clustering, because it is assembled without regard for any relationships among the characters in it.

The Marvel Universe is somewhere in between. It *is* more clustered than a random network would be, but not much more, and not nearly as much as other, real networks. What does this mean? No one seems to be sure at this point. The authors had this to say about their discovery: "It seems that Marvel writers have not assigned characters to books in the same way as natural interactions would have done it." On the other hand the Marvel Universe mimics its real counterparts in several ways, suggesting that the writers who invented it have, consciously or not, constructed a social world that is, in many ways, real. Much more real than the characters in it.

Does the Marvel network tell us anything about ourselves and the social networks to which we are attached? It's reassuring on one level: if you fear that, *Matrix*-like, we are all just playing out our lives under the control of some supreme beings or machines, you'd think that perhaps we could detect their creative hand by the absence of significant clustering in society. But we don't, so either we are truly free or our makers are better at creating societies than Stan Lee and his colleagues.

Do You Know 290 People?

W hether we are separated by a mere six degrees of separation—or even fewer—depends to a large extent on how many people we know. Take the extreme case: how many degrees are there between a hermit who lives deep in the New Brunswick woods and his counterpart at the northern end of Vancouver Island? The hermit has to go to the local store for supplies, so he knows the store manager. The manager in turn must know some local politician. Politicians of all stripes know each other, so it's probable that this particular politician knows the MP for the area, who in turn knows the MP whose riding includes northern Vancouver Island. The rest is just the reverse of what got us this far. But this is a slightly atypical example, a mix of types: a hermit is someone who minimizes the kind of personal contacts that make six degrees of separation possible, while politicians are anti-hermits.

What about more typical people? How many people do you know? This is not an easy question to answer, the difficulty being that tricky word "know." Does that mean, as it has in some studies, all the people that you know on a first-name basis, people that you could contact if you had to? Does it include people that you are likely to encounter in the next sixty days, or all the people still alive that you have ever known? That's one problem. Another is that if you were asked to write down the names of all those people, you would inevitably overlook some. Most people are more accurate about recent contacts (those with whom they've communicated over the last twenty-four hours rather than the last week), and they're least accurate about encounters occurring more than two weeks before. Anything more recent—or, surprisingly, more remote—is better remembered.

This difficulty is frustrating for social scientists, who would like to know how human society is divided up into networks of acquaintances. Understanding those networks would illuminate a long list of disparate social processes, from the spread of rumours to the transmission of disease. But how do you go about determining those networks in a way that minimizes the problems?

When you set this book down (I know, it's difficult), try it yourself. Write down all the people that you could recognize and address by name. Add the restriction that the person would also have to be able to address you by name—that eliminates movie stars, hockey players, rock musicians and politicians. But you're going to have to give yourself plenty of time. When this has been done in the past, people's estimates usually average out in the five-hundreds, but the variability is amazing. In one early study, office workers' estimates of the number of their acquaintances ranged from fifty to ten thousand. Their estimates *did* average 522, but it's hard to have much confidence in a single number like that when the guesses are all over the place. And that five hundred or so rises dramatically when people are reminded to include relatives, friends from high school, former colleagues, even members of clubs.

Just to get an idea of what sort of puzzle this is, it's worth taking a step back and comparing the number of people you know with the number of words you know. A common estimate for the number of words in the average North American's vocabulary is fifty thousand. (That number is usually reached by the mid- to late teens, which means that a typical person who starts talking at the age of two must acquire about fifty thousand words in fifteen years, or about ten new words *every day*.) But how long do you think it would take to list those words? Even if you kept a diary, adding the new words that you used that day, there are many that wouldn't appear even for months.

It seems that simply listing the names of your acquaintances might not be the best way of finding the total number, at least in any sort of reasonable time. Social scientists have tried a variety of techniques, and, as you will soon see, the number of people you appear to know depends on how you count them. One way to make sense out of all this is to start with techniques that yield high estimates, and work down from there.

A social scientist who was one of the first to be interested in this, Ithiel de Sola Pool of MIT, randomly selected thirty pages each from the Chicago and Manhattan telephone books, then scrutinized each page, looking for surnames shared by anyone he knew. He then multiplied the number of names by the number of pages in each phone book to get an estimate of how many names he would have collected had he scanned the entire book. Pool came up with 3100 people using the Chicago book and 4250 from Manhattan. Obviously these aren't very precise methods, but when you put them together it does look like Pool's acquaintances might have numbered in the high three-thousands.

This is a suspicious result, because Pool had tried something different earlier and had come up with a much smaller number. He had carried a book with him and recorded the name of every person he came in contact with. Pool had to exchange words with the person (even by phone), had to have met them before and had to know their name. He kept this diary for a hundred days. Of course the list grew quickly for the first few days as he encountered all the likely suspects, but the numbers started to dwindle as time went on. By days ninety-nine and one hundred, hardly anybody was added to the list. The total ended up being 685 people, and when Pool then extrapolated those hundred days out to twenty years, he came up with the conclusion that his personal "acquaintance volume"—the technical term for the number of people you know—was about 1500.

That seemed to be a close approximation to the number of people he really was in contact with, and it is a pretty healthy number of acquaintances, but it was still less than half the number derived from the telephone books. Why the difference? The telephone book allowed him to remember and recognize people whom he never came in contact with anymore but who qualified as acquaintances of his—if they happened to walk into his office he'd be able to say hello to them. But they weren't likely to do that, especially during the hundred-day period of the diary. You could argue that the telephone-book method gives better coverage of all the people you might know or have known.

Other researchers have since picked up on the telephone-book approach, although in every case have changed the technique slightly to avoid what were seen as obvious problems. For one thing, some names, like Smith, would likely take up an entire page, and if you randomly selected that page, you would instantly and significantly reduce the variety of names you were likely to see. On the other hand, most of us are likely to know at least one Smith, so you'd at least have one hit on that page. Whether these two opposing sources of error would balance is, as far as I know, unknown.

Linton Freeman and Claire Thompson in California attempted to eliminate these biases altogether by bypassing the pages and going right to the names in the telephone book. They randomly chose 305 surnames from the Orange County phone book and asked students who lived in Orange County to identify the names of acquaintances from that list.

After extrapolation to the entire phone book, the number of acquaintances per student ranged between 4707 and 6333, an average of 5520, not wildly different from Pool's original estimate of three thousand plus, but definitely bigger. Those numbers seem absolutely astounding to me, but at the same time hard to dispute, especially considering that, in this study, Orange County students used the Orange County phone book. The fact that different cities have different population mixes—and surnames—means that the best phone-book surveys should match the people, the place and the book.

However, in this case, the numbers turned out not to be as reliable as they seemed. The original Orange County study was published in 1989, but shortly afterward a different team of researchers went back over the Orange County phone book and found that small but significant errors had been introduced over the course of the study. Re-analysis lowered the number of acquaintances from an average of 5520 to less than half that, 2025. Again, that is likely a far larger number than the students would have been able to come up with had they used diaries or even just tried to list everyone they knew.

Pool's diary method has been tried with groups of people as well. One such study had people list their contacts for an identical period of one hundred days, with results that—again—were wildly inconsistent, anywhere between 72 and 1043. The point needs no belabouring: not only do different methods of estimating the number of people you know give different results but the same method applied to different people gives numbers that are all over the map.

Memory, or its failure, is at the root of the problem. Elaborate techniques that yield wildly disparate results, no matter how well designed, wouldn't be necessary if we could just sit down and come up with the names of all our acquaintances. But not only can we not do that, or are not willing to try, but most people have trouble remembering even what they have already remembered. One volunteer in a study recorded 117 acquaintances in five days, but nearly three years later could no longer remember or recognize 31 of them.

Telephone-book surveys yield the highest number of acquaintances, probably because they permit the inclusion of people from the past; diaries, though hard to maintain, reduce that number dramatically. But even they might be exaggerating the real world of acquaintance volume. Social scientists/mathematicians Russell Bernard and Peter Killworth and their colleagues have been grappling with this problem for decades, and they have developed methods of estimating the number of personal contacts that give answers in the low hundreds.

One in particular, called the "scale-up method," points to the number 290 as one of the best and most consistent estimates of the number of people anyone knows. Volunteers are asked to count the number of people they know who belong to a set of carefully defined groups: some examples from past studies have been Native Americans, postal workers, diabetics, prisoners, people named Christopher, Kimberley or Nicole, patients with AIDS, twins and even recent victims of homicide. The theory is that if one of these groups represents, let's say, 1 percent of the population, and a person knows three of them, then, at least on the basis of this one group, that individual likely knows three hundred people. It's inevitable that you will know more in some groups and fewer in others, but the more you're tested on, the closer you'll get to some sort of consistent number. As long as the actual number of such groups in the overall population is known, then volunteers' reported numbers can be scaled up to estimate how many people they know, period.

There are some potential problems with this method (what's new?). Participants might not be aware that some of the people they know belong to one of the subpopulations (like being a twin or being diabetic) or they might not all have the same opportunity to get to know members of those groups. An example might be knowing someone with AIDS—in any group of volunteers there will inevitably be some who are more likely to know people with AIDS than others. But there are ways of checking the numbers produced by surveys like this, and they are pretty consistent. And remarkably, several surveys of this kind have yielded the number 290, or something very close to it. There are exceptions: whereas randomly chosen people seem to know about 290 people, special groups, like clergy, know hundreds more (not really surprising given that clergy are obliged by the nature of the job to maintain large personal networks). But Bernard and Killworth and their co-workers think that so far, 290 is the best guess for a minimum number of people that we all know.

There have already been some intriguing, though depressing, applications of the work. Bernard and Killworth have used their scale-up method to answer the question "How many Americans either knew someone who died in the attacks of September 11 or knew someone who knew someone who died?" Their analysis suggested that just about 100 million Americans knew someone who in turn knew someone who was killed on that day.

That's where the research stands today: depending on the method you use, you can create lists of acquaintances that range from 290 or so up to a few thousand. Which is right? None or all of them: there is no "right answer"—the answer you get depends on how you've asked the question. In turn that is dictated by how much time you have, what exactly you mean by "know" and how well the tricks you've used to get over the memory hurdle have worked. But even with the uncertainty, social scientists are convinced that this research will lead to new thinking about the transmission of disease (which in many instances depends on personal contacts), the spread of information, especially through the internet, and the very structure of societies. It seems odd, doesn't it, that what should be the single most important number a social animal could know—the number of other animals it knows—could be so elusive.

Are You Staring at Me?

Imagine you're driving your Chevy Malibu on the streets of Palo Alto, California, in the early 1970s. You cruise up to a red light, stop, and when you casually glance to the side you're looking right into the eyes of someone a metre or so away. And he does not look away. He fixes you with an unwavering stare. Unsettled, you take off the instant the light turns green. The perfect reaction! Unbeknownst to you, you have just taken part in a classic psychology experiment, and your reaction to being stared at is exactly what the experimenters were looking for.

The experiment was an attempt to document how people respond to being stared at. It is surprising to me that most of the targets of the staring simply fled. I'm pretty sure that replicating that experiment today would generate plenty of aggressive reactions: haven't threat and counter-threat become part of road etiquette? The experimenters were well aware that a fixed stare is used as a threat by a variety of apes, baboons and monkeys, but they also knew that in most of those cases, it is the dominant animal that does the staring, and the typical response is one of submission or withdrawal from the situation. So the traffic light set-up was perfect. The experimenters wanted to know if a stare was a threatening gesture and also what would happen if the person being targeted was temporarily unable to escape from the situation—held back by the red light. Would their response be out of proportion once they were released? If so, it would make the observation of such behaviour much easier. Rocketing across an intersection is easier to quantify than a subtle turning away of the head.

Here's how it worked: the experimenter rode a motor scooter (described in the article in the *Journal of Personality and Social Psychology* as "dilapidated") up to the red light and waited for a car to arrive in the adjacent lane. As soon as that happened, the experimenter turned and stared—unblinkingly—at the driver. The distance between them was only about a metre and a half. As soon as the light turned green, a stopwatch clicked on to record how long it took the car to get to the other side of the intersection. The same experimenter on the scooter also served as his or her own control by sometimes sitting at the light and not staring.

The results were clear-cut: on average, drivers that had been stared at took about 1.2 seconds less to get through the intersection. It may not sound like much, but time yourself when taking off from a red light and you'll see: there's no doubt these drivers were spooked.

If you're one of those people who might actually have owned a Chevy Malibu (you still do?!) you're no doubt thinking that the acceleration of the stared-at drivers could be attributed to the fact that they thought they were being challenged to a drag race across the intersection. To control for that, the experimenters substituted a pedestrian standing on the street corner next to the light. Everything else, including the distance between the two people, was the same. So were the results.

One of the really intriguing parts of this experiment, and one that illustrates just how tricky experimentation can be, came about when the researchers felt they had to establish that it was indeed the staring that was prompting people to stomp on the accelerator and not just the fact that somebody next to them was doing something strange. In other words, was staring just one of many weird behaviours that might also have driven people away?

To test this possibility, they had to design a version of the experiment that was exactly the same in every way except that the experimenter, rather than staring, was doing something unusual, or as they put it in the paper, "incongruous." This was tricky, because the behaviour, whatever it was, couldn't be distracting for other reasons. For instance, they decided against having a man cracking a whip on the street corner for the good reason that this might have seemed threatening, and eventually hit on the following: a student sat on the sidewalk with a hammer and proceeded to hit and pick at the sidewalk when a car pulled up to the light. Now that's incongruous. Nonetheless, the hammering behaviour did not prompt drivers to flee the scene.

Scientists try to control everything in an experiment except the one variable they're interested in (in this case, the time to cross the intersection), but that's often easier said than done. For instance, in this experiment there was no way of controlling when the car-to-be-stared-at arrived at the light after the scooter, with the result that the actual staring time varied wildly, from about three seconds to more than twenty. But that uncontrollable feature turned out to be one of the most interesting parts of the experiment: the amount of staring time had no effect on the drivers being stared at. Those who had been subjected to the minimum amount of staring took off just as quickly as those who'd endured a half a minute of it or more. The conclusion: there's a moment—and it arrives quickly, in a couple of seconds or so—when a look becomes a stare. Once that threshold is crossed, the relationship between two people changes.

This experiment is one of hundreds that have attempted to explain what is surely one of the most subtle yet complex of human interactions: eye contact. The irony is that because we do this all our waking hours it's hard to see how it can be complicated. It all sounds so familiar: men don't engage in as much eye contact as women; when you're talking to someone you usually don't look at them much until you're just about to finish what you're saying; sometimes you glance at your conversational partners to keep them from butting in when you're in the middle of a pause.

In these situations eye contact is linguistic. But it's also emotional. If a picture is worth a thousand words, meeting someone's gaze across a crowded room is worth paragraphs of idle blah-blah-blah "And what do you do?" Making eye contact usually means the start of something, sometimes trivial, sometimes not; sometimes good, sometimes bad. That's one reason waiters are so good at avoiding it.

As fascinating as all this is, there's one aspect of eye contact that is, more than all these examples, a foundation stone of human behaviour. It's the role that gaze plays in understanding what others are thinking. Knowing what's on someone else's mind, something psychologists call "theory of mind," seems to be an essential human skill.

One neuroscientist, Simon Baron-Cohen at Cambridge University in England, has suggested that the human brain comes equipped with a pre-packaged set of "modules"—think of them as software programs—that work together to allow us to get into someone else's mind. They are collectively called "the mind-reading model." Although there's not yet enough evidence to say that such modules exist, Baron-Cohen's idea neatly puts together the mental operations that are needed to be able to know what other people are thinking. The four modules are: the intentionality detector (ID), the eye-direction detector (EDD), the shared-attention mechanism (SAM) and the theory-of-mind mechanism (ToMM).

Here's how they would work. The intentionality detector allows us to figure out the intentions of other living things. If the dog is chasing the cat, we figure it wants to catch the cat. This is a pretty low-level, unsophisticated mental ability. The eye-direction detector finds the eyes in a face, figures out from the position of the pupils where the face is looking, then makes the assumption that the person behind the face can see whatever lies in that direction.

The shared-attention mechanism takes things one step further. Baron-Cohen estimates that it starts working somewhere around one year of age, and it allows us to recognize when we are looking at the same thing as someone else: a child at the table glances sideways at the chocolate cake, and you use your SAM to deduce that it is the cake, not the carrots, that he is looking at. A birthday party of two-year-olds is full of SAMs doing these calculations.

Finally the ToMM kicks in, but not until as late as two or three years of age. It is the full-up, expert system in this set of mental modules, capable of going beyond the others to read minds, and by so doing, recognize deceit, pretend and even play games of imagination with others. It allows us to know what someone else is thinking.

It's not clear that the human brain has four discrete areas that match these, but there is abundant evidence that the capacity to read what others are thinking is crucial to human social life, and, at least for sighted persons, this ability depends on reading another's eyes. And it starts earlier than you could have possibly imagined.

A study published in 2002 showed that, from birth, infants prefer to look at faces that are looking them in the eyes. These babies were presented with pairs of pictures of the same person, one looking straight ahead, one with the eyes averted. The babies both looked more often, and significantly longer, at the picture with the eyes looking back at them. The researchers then recorded the brainwave activity of a group of four- and five-month-olds. They were searching for brainwave patterns that appear when adults look at faces, and they found that those waves peaked highest in the babies' brains when they were confronted with faces looking directly at them. Obviously eye contact is something that the human brain is alert to very early in life . . . it may even be part of a brain circuit that is pre-wired to do exactly that.

It's not difficult for us, even when we're babies, to figure out the direction a pair of human eyes is looking, because the pupils are surrounded by white. The white of the eye is clearly visible on both sides, and when you can see an equal amount of white on both sides, those eyes are of course looking right at you. Head position plays an important role too: if someone's head is half-turned away from you but she is looking at you nonetheless (a "sidelong glance"), the pupils of her eyes will be to the side, with most or all of the white on the other. It makes no difference to you—you still have no trouble figuring out that she's looking at you—but it does imply that your brain must be sensitive to a wide variety of eyes left, right, up and down to recognize when you're the object of a stare.

Incidentally, the straight-on-stare signal of pupil in the middle of the eye provides at least part of the explanation for the eerie tendency of faces in portraits to follow you around the room. If you watch a real face that stares ahead as you move around it, the position of the pupil will shift with respect to the white of the eye as you move. Move far enough to one side and you won't be able to see the white on the other side of the eye at all. But because the image in a painting is flat, no matter where you view it from you'll still see white on both sides of the eyes, and your brain will make the automatic judgment that the person in the portrait is staring straight at you.

The white of the eye is called the "sclera," and when a pair of Japanese scientists compared human eyes with those of at least half of all primate species in the world—gorillas, chimps, baboons and monkeys—they found our eyes to be unique. We're the only ones that have a white sclera, and we show off more of that sclera than any other animals. The scientists speculated that while some animals might benefit from a dark sclera (predators would have trouble telling whether the animal is aware of them or not), humans needed the communicative capacity of the white of the eye more, and so might have evolved from having "gaze-camouflaged" eyes to "gaze-communicative" eyes. Who would have anticipated that the whites of the eyes are actually communication devices, as much as a smile or frown?

Brain imaging studies have illustrated just how powerful the brain's urge is to follow someone else's eyes. In one study an area of the brain that plays a key role in shifting our attention from one place to another became active when presented with faces whose eyes were averted. In other words, the subjects in this experiment who saw faces with eyes looking to one side or the other didn't make a conscious decision to follow their gaze; their brains, entirely on their own, activated neural mechanisms that would start the search for the object of attention. We all know that we look where someone is looking; what we don't realize is the degree to which that has become an automatic brain function, and the importance it has for our social lives.

As is the case with any automatic system, there are times when the eye-direction detector can be fooled. If people are shown pictures of faces where the colour contrast in the eyes is reversed—the pupil is white and the sclera is dark—they have a lot of trouble deciding which way those eyes are looking. Much of the time they will apparently mistake the now-darkened sclera for the pupil and judge the gaze to be in exactly the wrong direction, even though the "pupil" in this case is the wrong size and shape. This mistake persists even when they *know* that they are looking at negative images, suggesting that the automatic judgment of gaze direction is a powerful neural mechanism that defies correction by the thinking part of the brain. This may be Baron-Cohen's EDD (eye-direction detector) at work.

Our heavy reliance on the dark pupil/white sclera might explain why the cheapest, cheesiest and yet one of the most highly effective horror movie stunts is to have the evil one's eyes change contrast from dark to light. (That they revolve at the same time has something to do with it as well.) In the same vein, you might have thought that your discomfort socializing when you have a hangover was simply because you were feeling lousy. But if your eyes are bloodshot, people have trouble reading your intentions, because there's no pure-white background for the pupil to stand out against. There actually is something in the scientific literature called the "bloodshot illusion."

In cases like these, when the usual pupil/sclera contrast isn't available, you can always use the direction someone's head is pointing to infer the direction of their attention. That's certainly all we have to go on when we look at someone from a distance. I used to play football against a quarterback who had perfected the art of turning his head in one direction but looking out of the corner of his eye in another. Most of us trying to guess where he was going to throw headed off in the direction his head was pointing, and he usually made us look pretty foolish. But even when you can see the eyes in a face perfectly well, a slight tilt of the head in one direction or the other can subtly change the apparent direction of the eyes, so an EDD might not be enough in itself—it might have to be paired with a head-direction detector.

When you compare our abilities to take cues from the direction the eyes are looking you realize just how important that skill is and how it separates us from even closely related animals. I have a standard poodle, a smart breed—as dogs go. Dogs are much more attuned to humans than to any other animals, including their close relatives the wolves or even chimpanzees. They will follow the direction a human is pointing, and my dog certainly looks into my eyes, but that's about as far as it gets. He certainly can't follow my gaze to something on the other side of the room, and I'd be shocked if he knew what I was thinking. (On the other hand, I know he's thinking about food whether I can see his eyes or not.)

There have been many attempts to see whether monkeys or apes are capable of the same sort of gaze-reading as we are, and the general rule seems to be that all of them are capable of at least some of the steps toward "mind-reading," that apes are further along than monkeys, but that in the end, even chimps and gorillas have not yet demonstrated unequivocally that they can read what's going on in another ape's or a human experimenter's mind. They are clever though. There's a story told of a juvenile baboon male who beat up a youngster, who then screamed for help. The adults came running, and the juvenile immediately stared off into the distance, as if the threat had come from some predator in that direction, and so escaped retribution. Smart, but not quite human-smart.

One of the reasons it is so important to understand not just the role of eye contact in social life but its brain mechanics is that there are some disorders where the system isn't functioning properly, the most prominent of which is autism. Simon Baron-Cohen is also a prominent autism researcher. He (and many other investigators) are convinced that the social difficulties of autistic individuals are rooted in an inability to engage in eye contact. It doesn't end there: that inability to make eye contact leads to an impaired ability to "read minds" at least partly because these individuals are unable to interpret the direction of gaze of others.

Autistic individuals tend not to look where someone is pointing, and seldom point to objects or people themselves. There is evidence too that the problem extends from these obvious deficits to deeper problems, such as an inability to comprehend what's in someone else's mind, or even that someone else could have something different on *their* mind. Young children learn quickly that a face with the eyes pointed up and away from them, focused on nothing in particular, is "thinking," but autistic children of the same age are incapable of making that inference. Following Baron-Cohen's model, they apparently are flawed in their ability to detect eye direction, share attention and acquire a theory of mind.

One study in England tested sixteen thousand eighteen-month-old infants for their abilities to point, follow someone's gaze or play a pretend game; twelve of the sixteen thousand failed all three tests, and ten of those twelve were eventually diagnosed as being autistic. Scientists now are trying to pin down the part or parts of the brain that might be impaired in individuals like this.

What's fascinating about this is the mental edifice that is constructed from such simple building blocks. It begins with the brain following the direction of someone's eyes and builds to the human personality in all its richness, from the joy of imaginative play to the cruelty of deceit.

THE LOOK . . .

Anyone who has experienced the sensation of locking eyes across a room with someone *interesting* knows that mutual gaze can be a powerful force. How powerful? It must be the force behind the idea that it is possible to fall in love "at first sight." An attractive face can turn heads, but it's not nearly enough. However, when eyes lock, something dramatically different happens. Just how powerful this is may surprise you.

One example is a set of experiments conducted by psychologist Ekhard Hess when he was at the University of Chicago in the 1960s. Hess wanted to find out whether dilated pupils had any effect on a person looking at them. He presented male volunteers with a variety of pictures, one of which was an attractive woman. In fact she appeared twice in the set, once with her pupils retouched to be highly dilated, another time with her pupils normal size. He found that a significant proportion of the time the men judged the version with the dilated pupils to be more attractive, and also, *their* pupils dilated when looking at her. Although some of the men thought the woman was "more feminine" or "prettier" in the picture where her pupils were dilated, none of them were actually aware of the pupils themselves.

But why? Hess performed other experiments that showed that our pupils dilate if we're looking at something or someone interesting. In one set of experiments he had people who were hungry view images of random objects, including slices of very delicious looking cake. Whenever the hungry people saw food items like the cake, their pupils dilated. If they weren't hungry, the cake had no effect.

So dilated pupils signify interest. If a man then looks into the eyes of a woman, and her pupils are dilated, he senses that she is interested in him. So, flattered, albeit unconsciously, he returns that interest.

Hess pointed out that this was likely the reason that Italian and Spanish women in the Middle Ages used to squeeze drops of what they called "belladonna" (beautiful woman) from the leaves or berries of the plant of the same name into their eyes. The active ingredient in belladonna is atropine, a chemical that causes the pupils to dilate. A very dilute solution is used by ophthalmologists to dilate your eyes before examining them. Atropine's molecular structure mimics that of acetylcholine, a chemical messenger that conveys nerve impulses to muscles. Atropine plugs into acetylcholine's receptors and blocks them, preventing acetylcholine from gaining access to them, and so suppressing the normal cycle of widening and narrowing of the pupil.

Of course the size of the pupils is only one of many factors that determine the attractiveness of a face, although pinning down exactly what it is that determines attractiveness has not been an easy task. In centuries past, mathematics has played a key role: medieval artists divided faces into seven equal parts, one for the hair, two for the forehead, two more for the nose, one for the space between the nose and mouth and a final segment for the chin. In some cases those distances were linked to each other, so that the ratios of, say, 3:2 could be found in the distances between different sets of facial features. But these were top-down decisions influenced heavily by numerological ideas that had nothing to do with beauty. Today scientists use experiments to tease out what makes a face beautiful.

Some of the most interesting of those experiments have used the blending of facial images to create faces that seem more attractive. A typical example would be to take two faces, match them for size, then average the differences between bright and dark all over the face. Those areas of light and dark provide the shading and lines of contour that give a face its three-dimensionality and texture. When the lights and darks of two faces are blended in this way the distinctive shadows and features are smoothed away, creating a face that is intermediate between the two. Studies of this kind show that most of us think the more blended the face, the more attractive it is. This is not the same effect as using a soft-focus lens to eliminate blemishes or wrinkles, because the technique works on line drawings of faces as well.

However, it doesn't seem to be true that average is the same as beautiful. More elaborate versions of this technique have shown that if faces that were attractive to begin with are blended, they are still more attractive than ordinary faces that are blended. The ordinary (the not-quite-so-attractive) never catch up. But if average is the same as beautiful, they should. After enough averaging, everyone should just be stunning! These findings work with people from different cultures: something about a blended, not-very-different-from-anyone-else face is attractive. Why this is so isn't clear. Some scientists believe that it's all about evolution and having babies, and that somehow an average face communicates something about health and child-rearing ability that an idiosyncratic face might not, but at this point no one really knows.

Back to the eyes. From all this you'd expect that an average face with big pupils should be just spectacular. It now seems that while this might be true, those big pupils have to be looking right at you. In 2001 scientists at the Institute of Cognitive Neuroscience at University College, London, performed a brain imaging study that showed that eye contact directly influences our judgment of the beauty of a face. Subjects, both men and women, were shown a series of forty different faces in various poses. The same face was shown either with eyes looking straight ahead (to make eye contact) or with eyes averted. In some cases the head was angled slightly to one side, again with the two different directions of gaze. The subjects in the experiment looked at the faces while their brains were being imaged to detect areas of activity. They then rated the faces for their attractiveness.

The results were, at first glance, strange. An area in the brain called the ventral striatum did become more active when an attractive face was presented, but only if that face made eye contact. The same face looking away caused the ventral striatum to *reduce* its activity. So this is a brain area that reacts to an attractive face but only if that face is looking into the eyes of the viewer. What does this mean? It was already known that the ventral striatum plays a key role in linking pleasure centres, reward, emotion and action. These results suggested that it was also important for, in the experimenters' words, "evaluation of stimuli with relevance to social situations." In other words, if you see an attractive face, the reward centres of the brain start to fire up, but if that face is looking the other way, the ventral striatum shuts down, dampening those high expectations. Yet if the attractive face returns one's gaze, then the ventral striatum kicks in. Remember "love at first sight"? It's not enough for the face to be beautiful—it should be looking at you too.

The researchers went on to suggest (although this wasn't part of the experiment) that an unattractive face might produce the opposite effect. That is, if the unattractive face made eye contact, the striatum would go into standby mode, but if the unattractive face looked away, the striatum would fire away, as the researchers suggest, in "relief."

It's striking, isn't it, that a judgment of attractiveness, coupled with an unconscious measure of eye contact, could lead to such potentially dramatic events in a life. In the brain, it's all about the expectation—or not—of reward. Love? That's for romantics.

PSYCHIC
STARING

Can you tell when someone is staring at you from behind? Do you get the feelings described by a psychologist in 1898 as "a state of unpleasant tension or stiffness at the nape of the neck, sometimes accompanied by tingling, which gathers in volume and intensity until a movement which shall relieve it becomes inevitable"? Various studies have estimated that 70, 80 or even 90 percent of people surveyed believe that they can tell when someone is covertly staring at them. Belief is one thing, but is there any real evidence? The picture at this point is more than a little clouded.

Unfortunately, the experiments on psychic staring that have been done by the pros don't clear things up much. In general, those whom you could define as paranormalists generally get positive results from experiments while skeptics get negative results, and the whole issue has become a parade of claims and counterclaims, confused even further by the fact that each new experiment is never an exact replica of the one before. This allows fresh claims to be made, but it also opens the door to new critiques. The best that can be done in this case is to give a flavour of the controversy, not a clear-cut answer.

The description of what it's like to be stared at that I quoted in the first paragraph was written by E. B. Titchener of Cornell University. Titchener authored a report in the journal *Science* called "The 'Feeling of Being Stared At.'" He began with an observation that would apparently be true today: that a certain proportion of his students every year believe that they can tell when they are being stared at.

Titchener did some experiments to test these declared abilities, but unfortunately he did not report the results. However, he did offer a neat explanation of the phenomenon that stood it on its head. He began by arguing that we are all conscious of what is behind us, especially if we're at some event where people are sitting behind us. That awareness shows as we pat our hair repeatedly or glance over our shoulders. He reasoned that this general nervous activity would sometimes provoke an actual backward glance. Titchener then cleverly pointed out that if you were sitting behind this person, and caught the movement of the turning head out of the corner of your eye, you would likely turn to look directly at him or her. And what would the person see? You, staring at him or her. And what would the person assume? That you'd been staring from the very beginning. A smart bit of reasoning, but no real evidence to support it.

Fifteen years later J. E. Coover tackled the subject, and this time did publish the results of a study in which he gave students (chosen because they believed they could tell when they were being stared at) the chance to guess whether they were being stared at or not during a fifteen-second period. A box with a die in it was used to make the staring unpredictable: an odd number of spots dictated a stare. Ten students each made one hundred guesses, and at the end of the experiment they had been correct on 50.2 percent of their guesses, a number that was, as Coover himself noted, "an astonishing approximation" to chance. Flipping a coin probably wouldn't have come as close to fifty-fifty as these students did.

Coover also asked the students to rate the confidence they had in each answer. He used this data to answer the anticipated criticism that those guesses in which the students had no confidence, and were indeed wrong, shouldn't be lumped together with high-confidence guesses. It could be that the high confidence was a real measure of a powerful psychic process. Unfortunately not: the students scored just as well on "pure guesses" as they did on guesses made with the absolute highest confidence.

Coover's study, as straightforward as it was, illustrates some of the crucial problems experiments on psychic staring run into. The staring/nonstaring choices have to be perfectly random: there can't be any possible way that the person being stared at, the "staree," can detect a pattern and begin tailoring his or her answers to it. That issue of randomness is still contentious today, as are the statistical analyses. Coover was pretty excited about the 50.2 percent, but in 2000 a group at Middlesex University in England reassessed his data and decided that, while it was true that the students were no better than chance at guessing when they were not being stared at, they were able to tell, at least to a statistically significant degree, when they *were* the targets of a stare.

The psychic staring action was pretty quiet in the decades following Coover's 1915 paper, but it heated up again in the 1990s. Study after study was published in the *British Journal of Psychology*, *The Journal of Parapsychology* and the *Journal of the Society for Psychical Research*. Some showed that psychic staring actually happens, some claimed to have shown that it doesn't exist. There's no point reviewing the results of the individual studies, although it is worth pointing out that with each new study novel methods were used to try to eliminate the possibility of cheating, conscious or not. So at first the starer sat right behind the staree; then the starer sat behind a one-way mirror, to eliminate the possibility of subtle sound clues created when the starer looked either at or away from the staree. The next step was to remove the starer to a separate room 20 metres away, connected only by closed-circuit video.

A dramatic technological step was then taken by William Braud and his colleagues at the Institute for Transpersonal Psychology in Palo Alto, California. They were concerned that the ability to know when you're being stared at depends on subtle internal signals, which could easily be swamped when volunteers were asked to guess whether they were being stared at. In that situation they are actively thinking about and attending to the kind of signals they're expecting to feel, and that sort of mental activity could blind the person to the weak body signals they might otherwise be receiving. If that were true, it would explain why the positive results that had been obtained to date were so marginal. So Braud suggested a radical rethink of the experiments: instead of asking people to say when they were being stared at, he devised a way of tapping into those presumed internal signals by measuring something called spontaneous phasic skin resistance response. This provides a way of recording unconscious nervous system arousal, like a polygraph (but, one hopes, with more reliable results).

Braud's set-up illustrates just how far experimenters have come in an effort to produce clean results, good statistics and no apparent opportunity for cheating. After the subject was shown the room with the television monitor on which he or she would be seen by the starer, the subject moved to a different room and sat in a recliner chair. Two electrodes for the SSR recording were attached to the left hand and the subject was then asked to sit quietly for twenty minutes, during which the starer was occasionally going to look intently at the subject's image on the monitor. The subject was asked not to guess when these stares might be happening.

Having returned to the original room, the experimenter/starer took "from a hidden location a sealed opaque envelope" that contained the random sequence for ten staring and ten nonstaring episodes, each one lasting thirty seconds and each followed by a thirty-second rest period.

To stare, the experimenter swivelled her chair to face the TV monitor and stared for thirty seconds. In nonstaring episodes she remained turned away from the monitor. To ensure there wasn't some sort of staring leakage, reflective surfaces in the room had been covered up, so she wasn't able to glimpse the screen inadvertently. The starer never saw the results of the recordings of the subject's nervous system activity—nor did the subject.

The results were provocative to say the least. The electrodermal recordings showed that there was, in at least three of four experiments, a statistically significant association between being stared at and nervous system reactions. What was most interesting was the kind of reaction. In the first experiment, starees recorded heightened skin responses, the kind you might expect with alertness or, in the traditional description of being stared at, a feeling associated with the hairs standing up on the back of your neck. After that experiment, the starer, Donna Shafer, underwent "connectedness" training, in which she became more comfortable with the idea of staring at and being close to others. In subsequent experiments where she again played the role of starer, she reported feeling "more relaxed, positive and nonanxious," and the skin responses of her starees reversed their direction from the first experiment, as if the people being stared at became more relaxed as well when she was looking at them.

There were additional details thrown in, one of the most interesting of which was the correlation of people's responses with their scores on tests of anxiety and social distress, which showed that the higher the score, the stronger the electrodermal change in response to being stared at. Braud and his colleagues suggest a number of possible reasons for this, including the possibilities that such people are more vigilant, that they are needier of social interaction and so detect it more easily, or that they simply are more used to being on their own and so didn't find the conditions of the experiment so strange and were, therefore, able to respond forcefully.

Okay—you may by now have been lulled into thinking that this all fits together and makes sense. If a calm, relaxed, "connected" woman is staring at you, presumably warmly, there's a good chance you're going to feel relaxed, not disturbed. But remember: this woman is staring into a television monitor; the staree is sitting with a camera. They are in two separate rooms with the doors closed between them. If the staree is feeling calmer because the woman is staring at him, there's some sort of paranormal thing happening here. I don't think—although the authors didn't explicitly suggest *how* this might be working—that parapsychologists would argue that somehow her transmissions went backwards through the video link, or at least through the ether. She's staring not at the subjects but at a pixelated image of them. What would happen if in fact that image was taped, not live? What if the starer was actually at the West Edmonton Mall? Would this strange experiment still produce positive results?

The notion of how this might work is of course what gives pause to most scientists (and that's putting it oh so mildly). There's no mechanism consistent with existing science. Not that there haven't been suggestions, one in particular from the man who's given "psychic staring" most of its publicity: Rupert Sheldrake.

In his 1994 book *Seven Experiments That Could Change the World,* Sheldrake argued that the general public should get involved and do experiments in controversial areas of science because the topics were revolutionary enough (or too far out on the fringe) for scientists to take them seriously. So, for instance, you can go to his website—"Sheldrake Online" at www.sheldrake.org—download instructions for how you and your friends can do psychic staring experiments and then send your results to him to add to his world-wide database on the subject.

The problem of course is that such evidence is never going to sway skeptics—there are just too many opportunities for mistakes or cheating, conscious or unconscious. Even more unfortunate from my point of view is that this approach—plus Sheldrake's high profile—makes him a tempting target for skeptics. There has been a tendency for them to target him, when the Braud experiments—and others—are the ones that should be attracting attention.

Besides suggesting that you try this at home, Sheldrake had a suggestion for how psychic staring works. In *Seven Experiments* he wrote, "Vision reaches out from the body. As well as light coming into the eyes, seeing goes out through the eyes . . . Permit yourself to think that your perceptions of all the things you see around you are indeed *around* you. Our image of this page, for example, is just where it seems to be, in front of you. The idea is so staggeringly simple that it is hard to grasp." The final step, of course, is that if images are actually reaching out from our eyes to touch the objects we're looking at, maybe if those objects are people, they can feel that we're looking at them. Ah, an idea so simple that I am having trouble grasping it. Wait a minute. It's not that it's simple—it's just that there's no evidence for it. Not a shred. Mysterious vision rays shooting out from my eyes to the objects I'm looking at? I don't think so.

This isn't the first time the idea has been put forward that images could be projected from the eye, although Sheldrake is one of the only ones to do it in recent centuries, or even millennia. In the sixth century B.C. Pythagoras suggested that the eye sent out rays to the object being looked at. This idea was recycled more than once, surfacing again in the Middle Ages. Of course it's one thing to reach out and touch someone who's sitting right in front of you and quite another to do the same thing via video camera, as in William Braud's experiments. But Rupert Sheldrake isn't, and shouldn't be taken to be, the authority on how psychic staring might work. Most of those who have done the experiments refrain, at least in their accounts of those experiments, from speculating about how it might work, although you can assume that it's thought to be part of the mysterious "psi" (paranormal or psychic) force.

One last comment about one particularly intriguing experiment. Marilyn Schlitz is a believer in these phenomena; Richard Wiseman is a skeptic. Together they performed two psychic staring experiments, one at Wiseman's lab at the University of Hertfordshire, the other at Schlitz's lab in California. Each acted as starer in experiments that were otherwise exactly the same in every detail. The amazing thing was that in both experiments, Wiseman got no effect but Schlitz found a significant effect.

This has been seen in experiments before and is called the "experimenter effect." Some of the suggested reasons for disparate results like this have been: that some experimenters have stronger psi abilities or were able to elicit the psi abilities of their subjects better than others (obviously experimenters who believe in psi would be likely to get better results than skeptics); that one group of subjects happened to be better at psi than the other; that there was cheating, either to create an apparent effect or to deny it; or that there were some undiscovered flaws in the experiment's design that might have biased the results one way or the other. Note that these explanations fall into two very different categories: the prosaic (something went wrong with the experiment) or the absolutely sensational (psi exists).

How well do those possibilities fit the Schlitz-Wiseman experiment? Experimental flaws are always possible and often hard to detect, but they seem less likely here because both Schlitz and

Wiseman ran the experiment exactly the same way in the same place: only the results were different.

Cheating? It would have been difficult for the participants to cheat, and one cheater wouldn't have been enough—for the results to have been significant, several would have had to cheat. Could either Wiseman or Schlitz have cheated? Several steps were taken in the experimental design to preclude experimenter cheating, but more important, the penalty for being caught—utter discreditation—would surely have prevented that.

The remaining possibilities depend on the existence of psi abilities, either of the subjects or of the experimenters. It seems odd to me to explain contentious results in an experiment by invoking the very phenomenon that is on trial. If you're a psi skeptic, then it'll be hard to buy the argument that Marilyn Schlitz is better at eliciting psi abilities from people than Richard Wiseman is. Wiseman himself isn't sure what happened, although he thinks everything should be scrutinized; an example would be to check the computer software to ensure that it wasn't being run differently in the two situations.

Am I losing my grip, or is this a truly significant experiment? As Schlitz herself has said, "This said to me that there was something important about the intentionality of the experimenter and something wrong with the whole notion of objectivity and detachment." No wonder most scientists shudder at the possibility that there might be something to these experiments: they chew away at the very basis of science.

As far as I'm concerned you can ignore all the believers who invite you to send in your results on psychic staring performed in your grandmother's kitchen, and the skeptics who sit down behind you in the cafeteria, stare at you and call that an experiment. I'm waiting to see what comes out of the investigation into Schlitz and Wiseman's.

A Study
in Scarlet

There's a rule of nature that is worth remembering: when something looks straightforward, it isn't. When anything seems to be just happening, there's usually something going on behind it that hasn't yet been discovered. As Shrek said, "Ogres are like onions . . . they have layers." If that's true, then nature is the ultimate ogre.

A good example, and one of the most captivating natural events of the year in North America, is the turning of the leaves in the fall. The colours *are* fantastic, but why green leaves become orange, red and yellow is a question that has puzzled biologists for more than a century. In the last few years some intriguing, though still tentative, answers to this question have been put forward, and they lend dramatic new depth to this piece of biological theatre.

I know as I write this that there will be those who think that scientists who investigate the fall colours will come up with data and theories that detract from the magnificence of the spectacle, thus draining away the magic, in the same way that Keats accused Newton of "unweaving the rainbow" with his revelations of the refraction of sunlight. To me it is the exact opposite: I still get excited at the sight of a brilliant rainbow, or better still, a double rainbow, but I can also get a second hit from knowing something about the amazing physics behind it. Sometimes I am even reminded of the sour, antisocial but brilliant Newton setting up prisms in his little room at Cambridge. You can now have the same sort of experience with the fall colours.

Pick up any biology book written before 2000 and you will read that the colours of autumn leaves are the result of a process that's more about removal and breakdown than creation. Summer leaves are full of chlorophyll, the molecule that captures sunlight and converts that energy into new building materials for the tree. (Chlorophyll is contained within chloroplasts, minute leaf organs with a unique history. They were once free-living algae, which, very early in the history of life, invaded a primordial cell and took up residence. Many hundreds of millions of years later, the descendants of those algae capture solar energy for all green plants living today.)

As summer turns to fall in the northern hemisphere and the sun travels lower in the sky, the amount of solar energy available to leaves declines. It becomes more and more difficult for the leaf to maintain the complex molecular assembly lines that convert that solar energy into useful products, and at some point the whole thing becomes a losing proposition. For many trees—conifers being an exception— the best strategy is then to abandon photosynthesis (the production of new material from sunlight, water and carbon dioxide) until the spring, when the amount of sunlight has rebounded to the point where it is again worthwhile.

However, even though autumn leaves are no longer able to do their job, they are still a precious resource, full of chemicals that will be useful to the tree once it starts collecting solar energy again. So before dropping their leaves for the winter, trees collect the most useful materials from them, shunt those materials back into the twigs, and only then let the leaves go. This recycling process starts in August, long before we see any changes in the colours of leaves, but at a time when the trees begin to detect a decline in the intensity of sunlight and a shortening in the length of the day.

Chlorophyll is one of the materials that the tree salvages from its leaves. Each chlorophyll molecule contains four atoms of nitrogen, an element too valuable to squander. So chlorophyll molecules are dismantled and their nitrogen shipped back into the tree to be held until the following spring. Leaves are green because of their chlorophyll (it absorbs other colours of sunlight but reflects green) and as chlorophyll is depleted, other colours that have been dominated by it throughout the summer begin to be revealed. These include the yellows and oranges of molecules called carotenoids, the same chemicals that give yellow and orange vegetables their colour. The carotenoids were there all summer, helping to capture some of the wavelengths of sunlight missed by chlorophyll, but we never see them until they are unmasked by the chlorophyll's removal.

That unmasking explains many of the fall colours, like the yellows of ash, birch and poplar. But it says nothing about the brilliant reds and purples of maples and sumacs, and those shades have forced scientists to come up with completely new theories about leaf colours, some of them radical.

The source of the red has been well known for a long time: it is created by chemicals called anthocyanins. What's puzzling is that the anthocyanin was not there all the time waiting to be revealed by the removal of chlorophyll. It is actually newly minted, made in the leaves at the same time as the tree is preparing to drop them. (In fact, manufacture of anthocyanin, like the breakdown of chlorophyll and other preparations for autumn, begins before we see any apparent changes in the leaf at all.) But it is hard to make sense of the manufacture of anthocyanins—why should a tree bother making new chemicals in its leaves when it's already scrambling to withdraw and preserve the ones already there?

There have been some weird and wonderful theories about these anthocyanins. Some have suggested that they might be defending the leaves against attacks by insects or fungi, or that they might attract fruit-eating birds or increase tolerance to freezing. There are problems with each of these theories, including the fact that leaves are red for such a relatively short time that anti-fungal and anti-herbivore activity don't really make sense. Leaf warming hasn't been directly tested yet.

One of the most intriguing ideas was published by Italian scientist Marco Archetti, who suggested that the vivid colours of fall were actually part of a game being played between insects and trees, a game with life-or-death consequences. (This idea had originally been put forward by the late William Hamilton, but it fell to Archetti to elaborate it.) Archetti's suggestion was that in the ongoing battle between herbivorous insects and the plants they feed on, it might be worth it for plants (in this case trees) to advertise that they are healthy and robust and easily able to mount chemical defences that would resist infestation. If insects paid attention to such advertisements, they might be prompted to lay their eggs on some other, less resistant host. One way trees could advertise their strength would be to display brightly coloured leaves in the fall.

If this were true, insects that laid their eggs on trees in the fall (Archetti and Hamilton focused on aphids) would eyeball the brightness of fall colours and lay their eggs only on trees that were duller and less vivid. If the game were being played straight up, those would indeed be trees that were less capable of defending against the aphids. However, in this case advertising is exactly the right word, because in the natural world, as in ours, advertising could be false. What would prevent a less than robust tree from displaying bright colours anyway and fooling aphids into avoiding it? The answer is cost.

Although cost might not apply to the yellows and oranges in the leaves that are simply revealed by the withdrawal of chlorophyll, it would apply to anthocyanins. It takes energy to make them, and if they were being made to dissuade insects, the energy expenditure would be justified only if insects actually did leave those trees alone. But if all trees, weak or strong, were producing brilliant fall colours, there would be no choice for insects to make. They'd lay their eggs anywhere, and that would mean that weak trees would be making the effort to display bright leaves but would be preyed on just as much as their more robust neighbours. They would have made an energy investment (one that further depleted their weakened state) but received nothing in return, and that is the kind of losing strategy that lands a species in the evolutionary trash can.

If, on the other hand, being preyed on by aphids ended up costing weak trees less than false advertising, then they would abandon the effort to fake the insects out, and an equilibrium would settle in, with strong trees advertising their strength with bright colours, weaker trees displaying their true, weaker colours and insects making the appropriate choices.

It's a neat idea, that the trees with the brightest-coloured leaves in the autumn forest are the most robust, strutting their stuff to potential insect predators. The problem—and I think you'd agree it's pretty significant—is that there isn't much evidence for it. No one knows whether aphids make choices according to the brightest colours; no one knows whether more robust trees sport the brightest leaves; no one knows whether other insects besides aphids might be deterred. It's an idea, but at this point, not much more.

When all the theories are lined up, there's one that explains best why leaves would go to the trouble of making red anthocyanins when they're busy packing up for the winter. Some call it the "light screen" hypothesis. It sounds paradoxical because the idea is that this red pigment, anthocyanin, is made in autumn leaves to protect chlorophyll, the light-absorbing chemical, from *too much light.* Why does chlorophyll need protection when it is the natural world's primo light absorber? Why protect it at a time when the tree is breaking it down to salvage as much of it as possible?

There are answers to those tricky questions: that's what I meant by nature having layers. For one thing, chlorophyll, although exquisitely evolved to capture the energy of sunlight, can in some circumstances be overwhelmed by it. Drought, low temperatures and nutrient deficiencies can all stress the photosynthetic machinery of the leaf and lead to its overexcitement if it is drenched with sunlight, and autumn leaves have to put up with low temperatures at least, if not the others. But the problem of oversensitivity to light is even more acute in the fall, because the leaf is busy preparing for winter by dismantling the machinery in it. This means that the energy absorbed by chlorophyll molecules when bombarded by particles of light isn't immediately channelled into useful processes and products, as it would be in an intact summer leaf. In fact, excited chlorophyll molecules in these unstable fall leaves can create forms of oxygen that would be extremely destructive to the leaf. Overall it's in the autumn leaf's best interests not to let its chlorophyll absorb too much light, and the solution is to make anthocyanins that block the light and protect the chlorophyll.

Even if you had never suspected that this is what was going on when leaves turn red in the fall, there are clues out there. One is straightforward: on many trees, the leaves that are the reddest are those on the sunlit side of the tree. Not only that, but the red is brighter on the upper side of the leaf. It has also been recognized for decades that the best conditions for bright red colours are warm, sunny, dry days and cool nights, conditions that nicely match those that make leaves susceptible to excess light. And finally, not that this is quite so obvious unless you do a lot of north–south travelling, trees such as maples are usually much redder—they produce much more anthocyanin—the farther north they grow. It's colder there, they're more stressed, their chlorophyll is more sensitive and it needs more sunblock.

On the surface, the red forests of September and October are attractive, sometimes breathtakingly so, but the real story is that they are running a race against time, under pressure to protect and remove as much of the valuable material left in the leaf before the first frosts shut them down completely. Photosynthesis generates the energy necessary to do this, and at this point in the season it is in a precariously vulnerable state, with the heart of its machinery, chlorophyll, being one of the primary leaf materials targeted for salvage. It is all a delicate balance: the leaf is likely using most of the energy it is still able to get from photosynthesis to shunt its most valuable materials back into the tree, at the same time siphoning off some of that energy to make anthocyanins to protect the dwindling reserves of chlorophyll from too much sunlight. If it all works out, the energy runs out at about the same time as the last chemicals of value leave.

This scientific story is still being written: not every detail has been confirmed by experiments, and it is still not clear why some trees resort to producing red pigments while others don't bother. Presumably trees that turn yellow—or don't turn any colour at all— have other means at their disposal to prevent overexposure to light in autumn. Their story, though not as spectacular to the eye, will surely turn out to be as subtle and complex.

The Ultimate Bargain Flight

W e are fascinated with all kinds of flight, from the dramatic liftoff of the space shuttle to the acrobatics of the hummingbird. Spectacular as both are, they are also enormously expensive, whether measured by dollars or calories. We fixate on speed and power, but that is a shortsighted approach. Even a very minimalist approach to flight, when examined closely, reveals the beauty of aerodynamics on the cheap. Some organisms are the equivalent of the small-market sports team—they have to make do with less. Maple trees are the perfect example.

No, they don't fly, but their seeds do. Maple seeds come encased in a thin fibrous layer shaped like a wing: the familiar maple key. The key is actually the maple's fruit, with the seed wrapped inside. As anyone who's lived near a maple knows, a stiff breeze at the right time of year will launch hundreds of maple keys, helicoptering down to the ground. It's that whirling motion, "autorotation" as the experts call it, that is a marvel of low-budget aerodynamics. The goal for the maple is to distribute its seeds as widely as possible. The slower they fall, the longer they're aloft and the more likely the wind will carry them out from under the shade of the parent tree. That's why playing the aerodynamics is so important for a maple (and for any other species that produces winged seeds). The keys don't look like much when you pick up one of hundreds left stranded in the corner of a parking lot, or in the gutter, but they are subtle enough in their use of the air that generations of scientists have been forced to dig deep for creative approaches to understanding how they work.

A chapter could be written about those experiments, from efforts in the 1800s to determine how maple keys fly by examining the marks they leave when landing in sand, to twentieth-century concatenations of lights, cameras and mirrors to record their flight instant by instant, to the design and manufacture of artificial keys, some of them up to 30 centimetres long, to experimenters plodding across wintry landscapes in snowshoes, tracking down pink-painted maple keys as they skid across the snow. But I'll just give you one example as a taste of the ingenuity needed to understand something as common, yet mysterious, as the flight of the maple key.

One of the earliest attempts to conduct experiments on winged seeds of any kind was described in 1933 by Howard Siggins of the California Forest and Range Experiment Station. Siggins was determined to figure out a way of recording just what happened when conifer seeds are released by a tree en masse; he wanted to know what kinds of seeds fell faster than others. But he couldn't just stand out in the forest and watch: too many seeds, too many gusts of wind, just too complicated. He had to find a way of watching seeds fall in a controlled situation.

His first attempt was to let them go from a height of more than 30 metres up a flagpole on the military parade ground at the University of California, Berkeley. Describing the results as "highly variable" (it turned out there were too many disruptive currents caused by warm air rising from patches of bare ground where students had been drilling), Siggins then had a moment of inspiration.

The bell tower at the university contains an elevator shaft, which runs an uninterrupted 60 metres from top to bottom. With the doors on all the floors closed, the shaft was a well-insulated chamber of still air, perfect for dropping seeds. But Siggins still had the problem of recording which seeds—redwood, Douglas fir or Ponderosa pine—fell fastest.

So here's what he did. When the seeds fell from a trap near the top of the shaft, they landed on a metre-square screen made of muslin (cotton gauze). One of Siggins' assistants had the job of over-laying a second screen six seconds after the first seeds landed, then another after six more seconds and so on (a second assistant cued him with a stopwatch). Eventually a stack of screens accumulated, each containing a cohort of seeds that fell at approximately the same rate. Labour-intensive though this experiment was, it was pretty neat.

Siggins was able to show that the seeds ranged widely in their rate of fall, from nearly 10 feet per second to just over 2.5 feet per second, and curiously that there seemed to be a tendency for the lighter seeds to fall faster. Of course Galileo is supposed to have proven that objects of different weights all fall at the same speed, but his Leaning Tower of Pisa experiment ignored the effects of air resistance, and stones aren't exactly aerodynamic. (It's also true that in all likelihood Galileo didn't actually perform the legendary experiment, at least from the tower, but it still makes the point.) Seeds can't fly but they are a lot better than stones. Something must have been controlling the rate of descent, and remember, for a seed, the slower the fall, the better. But while Siggins suspected that better aerodynamic performance had something to do with the length of the wing of the seed and the circle it swept out in the air as it fell, he didn't do any further experiments to prove that. Mortality stood in the way: Siggins died even before his report was published.

The vast majority of research on the flight of winged seeds has been done since the 1970s. Much of it involves formidable aerodynamics, but the subject can be approached in a slightly more user-friendly way by just picking up a maple key and examining it. The technical term for a maple key is a "samara," an elegant term that I'll use from here on in.

Samaras differ from species to species, but in general here are some of the things you'll notice the next time you sweep them off your windshield. Try balancing the samara on the tip of your finger. You'll find that almost all of the wing of the samara is sticking out on one side: most of the mass is concentrated in the seed. That turns out to be an important factor in its flight. Second, check out the shape. From above you'll note that the samara is narrow at the seed end, then broadens gradually, finally tapering a little as it nears the tip. Finally, run your fingertip along the surface. It's roughened, and thick with veins. Most of these veins are concentrated toward one side of the wing, the side that turns out to be the leading edge that cuts into the air as the samara turns.

Those brief observations are your entree into the world of samara aerodynamics, or how to get the most out of flight for the least. So toss that samara into the air: here's what happens as it falls.

The samara rises until gravity starts to accelerate it earthward. Now it's in the position it usually is when it has broken free of the tree. (David Greene of Concordia University has shown that samaras are more likely to break free on dry, windy days—the low humidity promotes drying of the attachment to the twig, and the wind provides the drag forces that actually snap that attachment, causing the samara to break away. How perfect that it falls in a strong wind that's likely to carry it away to greener fields.) It starts to plummet straight down, but only for an instant. Once the tip tilts even a minute amount from the vertical, the wing surface catches the air, and the samara starts to rotate—all within a few hundredths of a second.

Once spinning, most samaras drop at about a metre per second. They're then in equilibrium: gravity is pulling them down (and wanting to accelerate them downward) but the wing is counteracting gravity by generating lift; that is, it's pushing down on the column of air it's travelling through, and that slows its descent. In some ways the samara is like the blade of a helicopter: the faster it rotates and the larger the circle it sweeps out, the greater the lift. Neither is close to the aerodynamic ideal, a solid disc pushing down on the air (like a Frisbee dropped from the hand), or even better, a parachute. In their struggle to model this complicated behaviour, aerodynamics experts dreamed up something they call the "ideal rotor," a device that is infinitely thin and is made up of an infinite number of blades, and then compare everything else to this. A helicopter blade isn't much of an ideal rotor because it covers an area only about 5 percent of an entire disc, and a samara isn't much more, about 12 percent. But a samara does pretty well when airborne anyway. It sinks about 25 percent faster than a flat disc, 38 percent faster than a parachute (of the same mass) and about the same as a bird . . . that isn't flapping its wings. However, it doesn't have to worry about stability in the air the way these other flying objects do.

The larger the area that a helicopter blade or a samara sweeps out, the closer it is to aerodynamic perfection. That's one way the samara benefits by having the seed at one end. Any object of this shape rotates around a point that's very close to its centre of mass, and the closer that concentration of mass is to one end, the greater the area swept out by the wing. It's the difference between a cheerleader holding the twirling baton in the centre or at the end. For good performance in a flying object like a samara, that centre should be within 30 percent of the end of the wing, and maple keys are well within that limit. So positioning the seed at one end allows the samara wing to sweep out the largest area possible and provide the greatest lift.

What about the shape, thin at the seed end, gradually broadening, then slimming down again right near the tip? The more wing surface area, the better, again because the greater the area of the wing, the slower the rate of descent. But simply shaping it like the blade of a fan wouldn't be good enough, because there are subtleties of airflow at work here.

As the samara falls through the air, the speed of the air flowing over its surface varies: the farther out toward the tip from the seed, the faster the air is rushing over the surface, simply because of its distance from the centre of rotation (the horses at the perimeter of the merry-go-round go faster than those near the hub). To take advantage of this higher-velocity air, the blade should be wider. On the other hand, there are problems at the very tip. A broad, blunt end would encourage the development of little vortexes of air swirling around the tip, creating drag, and slowing the samara's rotation. It's much better to build a samara that slims down at the tip and use the material that's been saved to make the whole thing a little longer (and by doing so give it a little more lift), which of course seems to be exactly what has happened.

One of the most dramatic differences between a maple key and a helicopter blade or an airplane propeller is the texture of the surface: blades and propellers are smooth, samaras are not. In this sense samaras are much more like the wings of insects or small birds. This all makes sense to aerodynamicists because the forces that are most effective in permitting flight are different for different-sized objects: the challenges of getting a 747 into the air are completely different from those involved in delaying the fall of a maple key. A rough texture creates turbulence right at the surface of the key, and that turbulence helps decelerate the fall. The dimples on a golf ball create the same sort of local turbulence and dramatically increase the amount of time the ball stays in the air.

The veins that run along the samara's surface contribute to surface roughness but also play an important role in flight stability. You'll notice that most of them cluster toward the leading edge of the samara, the edge that cuts into the air as it falls. They help shift the weight of the wing forward toward that leading edge—the back edge has fewer veins and is lighter—and that keeps the falling samara stable, and hence airborne longer.

The laws of aerodynamics dictate that most of the weight at any point along the wing should be closer to the leading edge to maintain what's called the "angle of attack," the tilt of the leading edge of the wing as it bites into the air. The clustering of the veins ensures this proper weight distribution, and samaras without such veins instead have a thickened leading edge. The veins also prevent the blade from becoming distorted as it dries out, a good thing because a drier, lighter samara flies farther.

All of these features are easily overlooked and yet absolutely critical to the point of the whole exercise: stay in the air longer and get as far away from the parent tree as possible. There are other features of these structures that I haven't mentioned that play important roles too. The truly beautiful thing about a falling samara is that all aspects of its shape and distribution of weight go together to provide a package of self-correcting measures when things go wrong. Any in-flight failure, such as tilting up too steeply and starting to stall, or pitching too far forward and losing altitude, immediately changes the direction and pressure of the air flowing over the surface, which in turn restores stability.

Maples aren't the only trees that resort to equipping their seeds with wings in order to enhance their chances of survival. Others have experimented with slightly different designs, apparently under different pressures of survival. The tree of heaven, *Ailanthus*, a species that was introduced to North America and is now common in urban areas, produces a samara that both helicopters *and* rolls as it falls. On the surface this seems to have been a bad choice, because *Ailanthus* seeds fall more rapidly than maple keys. But they roll the same way a playing card does if you flip it into the air, and that rolling makes the card—or the seed—very stable in turbulent conditions. If the winds that liberate the seeds from the trees are strong, *Ailanthus* seeds might just gain an edge. Maple samaras, although beautifully equipped to fall slowly in moderate winds, can lose their equilibrium in very strong gusts and fall like stones.

It is, after all, all about trade-offs. A tree has only so much it can put into seed-making, and those limits constrain what it can do in the way of samara design. It likely wouldn't be a good idea to build one perfect samara every season—it might stay in the air for minutes and *still* land on the window of the next-door neighbour's SUV. So a tree makes the best seeds it can afford to with the materials and energy at hand, nicely but not perfectly designed, and sends them on their way. They don't fall like dead weights, but they don't soar like the birds either. Nature is always about making the best of an imperfect situation, yet somehow the results are never dull.

Catching
Flies

Playing catch can be as simple and unchallenging as tossing a tennis ball back and forth, but there is something satisfying about throwing, watching, tracking, catching and just generally wasting time. Playing catch must be one of the simplest repetitive activities that humans can indulge in (and we all know how enjoyable *they* can be), but you'd never know that from reading the psychological literature on the subject. Catching a ball—what some psychologists like to call the "outfielder problem"—is a subject of serious controversy.

Before I drag you into the depths of this contentious issue, there's an important point to be made. The outfielder problem is not actually a problem for outfielders; it's a problem for psychologists and a handful of physicists (some of whom are amateur outfielders too). Most people who learn how to catch a fly ball do so early in life, and never, ever need to know what's going on in their brains while they do it. But as psychologists have learned, it's not a trivial problem to figure out what we do when we watch a ball flying through the air, move to the appropriate spot, then catch it. The catch itself is tricky: as you might remember from your own experience, or from watching children learn, reaching for the ball at the final moment and closing fingers around it is usually the last part of the entire skill to be learned and likely involves some of the most complex brainwork. But the psychologists aren't there yet—not by a long shot. They're having trouble understanding the simplest part of that sequence: tracking the ball through the air and figuring out where it's going to land.

It was a physicist, Seville Chapman, who led off the modern discourse on catching a ball, with an article in the *American Journal of Physics* in 1968. Chapman selected the simplest situation of all—a ball hit straight at a fielder—and proposed a strategy that would allow the fielder to catch that ball. (Interestingly, although this scenario is a simple example from the physics point of view, most ball players agree that a ball hit straight at them is the most difficult to catch.) The best way to understand what Chapman was suggesting is to actually go outside and have someone hit a ball to you, but for the moment, try imagining yourself standing in the field waiting for that ball. Remember, it is coming straight at you, so as it leaves the bat and starts to rise, it appears to be climbing straight up in the air. If the ball is going to land in front of you, then from your point of view it will appear to rise, slow, then stop and fall straight to the ground.

Chapman argued that one way you could ensure that you would arrive at the place where the ball was going to hit the ground would be—at every moment—to position yourself so that the ball in flight appears to be rising at a constant rate, and *keep it that way* until the ball falls into your hand. That may sound impossible, because even though the ball appears to rise at first, it soon reaches its peak and starts to fall. But track a fly ball in your mind's eye and you will see that this is actually what you do.

Imagine again that a ball is hit straight at you but softly, so it will fall short of where you're standing. As soon as the ball appears to be slowing as it climbs, that means it is going to fall to the ground in front of you, and you'd have to run forward. If you did, you'd restore the appearance that the ball is climbing. If on the other hand the ball was hit well over your head, you would have to retreat to slow the apparent climb of the ball. Either way, if you're fast enough, you should arrive at the right place at the right time. The key is to avoid acceleration: as long as the ball appears to be moving steadily, stay where you are; if it starts to slow down or speed up, you have to move. It's a case of cancelling the ball's apparent acceleration whenever it appears.

Some researchers have even likened the appearance of a ball hit directly at you to the image of an elevator rising steadily in a tall building located at home plate. If the ball were superimposed over the elevator, all you would have to do is move so that the ball stayed in place in front of the elevator, and it would fall into your hands. (It sounds weird, but it works.) On the other hand, if a ball is going to fall in front of you, the imaginary building is tilted away from you, and you have to move forward to make the ball appear to keep pace with the elevator. And a ball that's going over your head would be keeping pace with an elevator climbing a building tilted toward you.

Chapman's strategy is simple enough and works well for this rare circumstance of the ball coming directly at you, although, as is often the case in theorizing, some important bits of real life are omitted, like strong winds or even just plain air resistance. Some calculations show that a batted ball goes only 60 percent as far in air as it would in a vacuum. That shortened flight also has a lopsided shape—it isn't the perfect parabola that Chapman envisioned. But those are minor problems compared with this question: what if the ball is moving to one side or the other? That appears to be a completely different story.

Chapman recognized that this case was a little trickier and suggested that this time the outfielder's brain had to keep two things in mind. One is the recipe I've just discussed: move so that the ball appears to climb steadily. The other is to try to maintain a constant compass direction to the ball. Imagine you are in centre field and the ball is hit to your right. You then have to run to make the ball appear to be rising steadily, and at the same time keep the ball, say, to your south-southwest, and once again, you'll end up in the right place. This special case, the "ball hit to the side" (Chapman referred to it as the "ball hit *somewhat* to the side"), seemed to him nothing more than an afterthought. He devoted a single paragraph to it in a three-page article—the second-last paragraph at that—but it has turned out to be the most contentious part of his theory. Chapman does get credit for being the first to attack the problem, but of course the pioneer inevitably becomes the target.

The psychologists who have followed up Chapman's work are convinced that Chapman had it wrong when he suggested we track balls on one side or the other by maintaining a constant direction to them. In fact, they're not even sure he got the simple ball-hit-straight-at-you version right either.

Some lab studies have suggested that people are not particularly good at judging the presence or absence of acceleration (as Chapman had required). Asking them to do that with the precision necessary to catch a ball seems—at least to some researchers—to be a tall order. On top of that, tracking the ball hit to one side necessitates keeping it at the same compass point as you approach it, adding a tricky second calculation to the mix. Yet most people who can catch a ball agree that it's easier, not harder, when the ball is hit to the side. Paradoxically, fly balls hit directly at you are the hardest to judge but the easiest to describe psychologically.

These difficulties prompted psychologists to go back to the lab to refine their thoughts on the outfielder problem. Out of this rethink came another theory. Where Chapman's is called optical acceleration cancellation (OAC), the new theory is LOT: linear optical trajectory. LOT suggests that when you're chasing a ball hit to one side or the other, you don't pay attention to the speed of the ball or its acceleration or deceleration but simply track its path against the sky, steadily moving to ensure that the path does not deviate from a straight line.

I can see that in my mind's eye as well: if the path of the ball appears to be bending, I accelerate or decelerate to straighten it out. This, according to the experts, is an easier job for the human brain than judging acceleration, the requirement for Chapman's OAC theory to be correct. But no matter what is proposed for the brainwork that goes into catching a fly, no matter how satisfactory it is geometrically, there still has to be a way of determining whether or not that's what people actually do. The tricky part is designing the right experiments.

One clue to what's going on in fielders' minds is to predict whether they'll close in on the ball differently if they're using the OAC mode rather than the LOT. In other words, as they run are they measuring the speed of the ball or its path across the sky? Michael McBeath and Dennis Shaffer at Arizona State University, the psychologists who first proposed LOT, suggested that fielders using that technique would likely accelerate first, then follow a curved path toward the ball's landing point, finally decelerating before the catch. On the other hand, fielders using OAC should run straight to the catching spot. McBeath and Shaffer videotaped outfielders as they caught fly balls and found that just over 70 percent of their running paths curved in a way that suggested they were using LOT, not OAC. When cameras were mounted on the fielders' shoulders, they revealed that the fielders were moving so as to make the ball appear as if it were travelling in a straight line, again supporting LOT.

These results were published in the journal *Science* in 1995. Satisfied that their data supported their model, McBeath and Shaffer (and co-author Mary Kaiser) added one last point at the end of their paper. We've all seen outfielders run smack into the outfield wall chasing a ball. According to the research team, that is because they are using LOT to determine that they *will* catch the ball; they just don't know when or where.

It might surprise you to know that the outfielder problem was not considered solved by the LOT theory. In fact, it became, if anything, more controversial. In response to the McBeath, Shaffer and Kaiser article in *Science,* a team of physicists at Purdue University (I don't think they were avenging their colleague Chapman) argued that LOT couldn't be right because they were able to concoct scenarios in which an outfielder using that exact strategy would end up more than 5 metres away from where the ball would land. The problem in their eyes was that even as a fielder is running to keep the ball's apparent path in the air straight, the ball could reach its maximum height, then start to fall, *all along the same straight line.* In other words, if you're concentrating only on keeping the ball's path straight, you might miss the fact that it's about to hit the ground.

McBeath's team acknowledged that difficulty by arguing that they never said that LOT was the *only* thing you needed to catch a ball hit to one side, but simply the primary method. Maybe, they added, outfielders might resort to a little OAC as well. In fact, as the ball gets closer and closer, it's quite likely that other perceptions come into play. In the last half-second or so, the ball is getting so close that it looks like it's getting bigger and bigger. That apparent "optical expansion" of the ball is likely to take over at the end to ensure that you're able to put your hand in front of the ball. Certainly if a fielder is forced to run hard and then make a diving catch, all concentration in those last moments must be devoted to making the hand meet the ball.

The outfielder problem likely involves several different brain tracking mechanisms; these probably include LOT, OAC and who knows how many others. Without wanting to belabour this, I'll just add that there have been some intriguing experimental stabs at the question. One actually dared to ask whether tracking the ball with the eyes was enough. While these experiments showed, not unexpectedly, that fielders follow the ball throughout its flight, they also revealed that a luminous ball can be caught—on the run—in a completely dark room. What does this mean? If you're going to track the ball against the background (as would certainly help if you're trying to judge either the ball's speed or its direction), you need a background! But a dark room doesn't provide one. These experiments suggested that your awareness of the movements of your head and neck as you track the ball are likely contributing to knowing how the ball is travelling and where it's going to land. The neural signals that allow us to know which direction our head and eyes are pointing are very fast, and very good at detecting accelerations of the head.

Some researchers have tried to simplify the situation by making the argument that all you really need to do is keep the ball somewhere between horizontal and vertical, between the ground and the overhead point, and you'll end up catching it, although the trickiness of this is to ensure that if it's climbing rapidly, but still not yet overhead, how do you determine that you'd better backpedal just in case? If you watch outfielders during a game, you'll see that they vary their approach from one fly ball to the next, yet both major theories demand greater consistency than that. It's also true that sometimes the game situation dictates how a fly ball should be approached. If there's a runner who will try to advance from second to third as soon as the ball is caught, a good fielder will set up slightly behind the catching point so he or she can move forward to catch the ball and then use that forward momentum to add power to the throw to third. There is no accounting for that in these theories.

There are occasions in baseball where a fielder seems not to follow any of the above theories. The best example is likely Willie Mays's amazing catch in centre field in the 1954 World Series. Vic Wertz of the Cleveland Indians smashed a ball to the deepest part of the Polo Grounds, a hit that in almost any other park—or in any other part of this park—would have been a home run. Mays took off at the crack of the bat and ran full out straight away from home plate, and was able to catch the ball only by reaching out at the last second. The amazing thing was that at most Mays glanced once or twice at the flight of the ball during his run; it was as if he had known immediately where the ball was going to land. But that's not supposed to be possible—at least for humans. There is, however, a fish that can do that: the archer fish.

This is one curious beast. The archer fish seeks prey that lives out of the water. It keeps an eye on overhanging vegetation, and when an insect alights, the archer spits a spray of water drops at the insect, more often than not knocking it into the water. This is a wondrous performance in and of itself, at least partly because the fish has to deal with the refraction of light where the water meets the air. We all know that if you look at a stick that is half in and half out of the water, it appears to bend. If you reach for the image of the end of the stick, you'll miss the stick itself. The archer fish somehow deals efficiently with the same problem (and not by aiming only at insects directly above it, as several documents on the internet suggest) and brings down its prey with regularity.

But knocking an insect into the water is only the first step. There are usually other archer fish spectators, and once an insect is dislodged, it's a race to see who will get to consume it. This is where the archer fish's remarkable fielding abilities come into play. Experiments performed by scientist Stephan Schuster at the

Albert-Ludwigs-Universität Freiburg show that within one-tenth of a second after hitting its prey, the archer fish shooter and its companions take off toward the spot on the water where the insect is going to land. The amazing thing is that after that point, the fish never needs to look at the insect again—in that first brief moment it has calculated everything it needs to know, and unlike baseball players, it never needs to track the object of interest—it simply heads for the impact point.

Schuster devised some clever experiments to prove the point that the archer fish makes up what passes for its mind in a fraction of a second. First he tethered a fly to a thread, then allowed an archer fish in an aquarium to shoot the fly down. The problem was the fly dropped only halfway toward the water, then hung there from the thread, swinging back and forth. The fish, oblivious to the fate of the fly, continued to the place where it expected the insect to fall. That showed that the archer doesn't follow the prey all the way but it still allowed for a certain amount of early tracking. The question then was, is the fish actually taking a mental snapshot of the fly's early trajectory, then following that out to where it would meet the water? Or was it doing something much simpler: just noting the speed and direction and inventing a trajectory to follow? So Schuster went one step further, and placed a sheet of clear glass over the aquarium. This time, when a fish shot at an insect on the surface of the glass, Schuster simply directed a blast of air at the fly so that it would skid across the glass in roughly the direction it would have headed had it been free to fall. Even though the fly was travelling horizontally, not downward, the fish still went to the same spot on the water, waiting there like Captain Hook's crocodile.

These experiments point to a general behavioural truth: the simpler the method the brain uses to accomplish something, the likelier it can be fooled. The archer fish uses much less brain power to catch a fly than we do, but is much more prone to error. However, those errors are so infrequent in their natural environment (there aren't many insects that are tethered to the branches on which they sit) that it's worth it for them to use "tracking lite." Humans, on the other hand, have million-dollar contracts to live up to, and Gold Gloves to be won, so LOT and OAC are the way to go.

SKIPPING STONES

Culture and technology alter daily life at a sometimes bewildering rate, but happily, some experiences stubbornly defy change. The child-like joy of skipping stones is one of them. Here's the first sentence from a scientific paper on the subject written in the 1750s (!): "It is very well known that stones thrown slantwise upon water skip, and this is a favourite amusement of boys playing on the banks of rivers."

And this is also the first sentence from a scientific paper on the subject, but this was written in 2002: "Nearly everyone has tried to throw a stone on a lake and count the number of bounces the stone was able to make. Of course, the more the better."

Nothing, it appears, has changed, except the suggestion that skipping stones is something only boys might want to do. Yet despite the ongoing fascination of the skipped stone, the understanding of what is actually happening has not moved forward any significant distance. In fact, the thoughts and observations of the scientist whose 1750s quote appears above are pretty much in line with what we know today.

That scientist was Lazzaro Spallanzani, and if this were anything but a book on the science of everyday life, I would devote this chapter entirely to him, not to the science of skipping stones, because his was a truly extraordinary scientific mind.

Just to set the scene for his work on stone skipping: Spallanzani is best remembered for his experiments demonstrating that life could not just arise out of thin air. The theory that it could was called "spontaneous generation," and there were some common observations that seemed to support it: leave some hamburger (or the eighteenth-century equivalent) on a countertop and pretty soon it would be swarming with maggots. Put some old rags and grain in a sack and a week or two later you'd generate mice. It appeared as if these inanimate materials contained some sort of life force that, given the right conditions, asserted itself and created living things.

When Spallanzani turned his attention to the problem, there had already been demonstrations that cast doubt on the idea (if you cover the hamburger so that flies have no access to it, you get no maggots), but there was still a widespread belief that life could be produced from non-life. He refuted the theory by putting broth, boiled or not, into a series of flasks. He sealed some of the flasks but left others loosely corked. Spallanzani was sure that the micro-organisms that appeared in open flasks of broth came from the air, not the liquid, and his experiment vindicated him: there was no growth in those flasks that had been boiled for a long time and subsequently sealed. Inadequate boiling permitted growth, as did loose corks. Even though the idea of spontaneous generation still hung on until Louis Pasteur's time, Spallanzani's experiments helped turn the tide against it.

But Spallanzani did much more than this. He was the first biologist to show that in the absence of sperm, fertilization would not occur, he performed a series of experiments that determined that bats navigate using their ears, not their eyes (see the chapter "Echolocation—Our Sixth Sense?") *and* he studied stone skipping.

Spallanzani's dissertation on stones and how they skip is one of the lesser-known pieces of his research. After a long search on the internet I was finally able to find a partial copy in the public library in Bologna, Italy, by keying in the Latin title: *De Lapidibus Ab Aqua Resilientibus.* It may not be well known, but it is a beautifully written article that explores some of the same features of stone skipping that still interest scientists two hundred years later.

Spallanzani begins by considering the idea, put forward by others, that it is the elasticity of water that causes the stone to bounce off the surface. The fact that he devotes several pages to exploring this notion is a reminder that even decades after the alpha male of physics, Isaac Newton, left his mark on the field, the natures of matter, energy and force were still uncertain and even mysterious. Spallanzani goes to some lengths to show that the elasticity of water—whatever that actually is—has little to do with the skipping of a stone. But it is when he gets to the point where he is actually throwing stones into water that his experiments begin to take on a very modern appearance.

For instance, he points out that if you drop thin, flat stones into water point first, they sink (as he put it, they "sought the bottom"). However, when he threw them with "the broad face of them turned to the surface," they skipped. Any stone skipper today knows this, and I suspect the same was true in the 1700s. But Spallanzani then went further and pointed out that stones will skip *higher* if they are thrown with a slight upward tilt to the leading edge. What's more, he noted that even when such a stone hit the water it remained tilted, and continued to stay in that position throughout the subsequent skips. Then he made what he apparently considered a crucial discovery: rather than focusing on the orientation of the stone, Spallanzani examined the effect of the stone's impact on the water. As he described it, "as soon as it was struck by the stone, it was depressed into a hollow consisting as it were of a double inclined plane, and the stone was seen to descend by one plane and ascend by the other, then to spring off." The man must have had sharp eyes to see that a skipping stone slides down one side of a depression in the water's surface, then up the other and off again into the air. As you will soon see, it took until 1968 and the use of high-speed film to confirm his observations.

Spallanzani was even able to launch a stone into the air by grasping it between his thumb and forefinger, brushing it across the surface of the water for about a hand-width, then letting it go. He noted that the bow wave up which the stone rides when it takes off was clearly visible. This brief description doesn't do justice to Spallanzani's thoroughness: he went on to fire lead shot across the surface of a body of water, then into clay, to compare the angles necessary for a skip. You might think bullets don't have much to do with stones, but Spallanzani was trying to establish why it is that you can skip a perfectly flat stone standing upright, but if your stone is rounded, you have to bend down and throw it almost parallel to the surface of the water. You do that to reduce the angle the stone makes with the water at impact: the lower the angle, the more likely a skip. As Spallanzani pointed out, the larger the angle, the greater that part of the impact force that is directed downward into the water. If there's too much downward force, there's no skip.

Think of it this way: a thrown stone hits the water with the same force, regardless of its orientation at impact. If it hits flat side on, the force of impact is spread out over the water's surface, enough to ensure that the stone can rebound. Hitting edge-on, on the other hand, ensures that that same force is now concentrated on a thin edge—that, plus other factors such as the shape of the stone, the angle with which it hits the water and the drag forces of the water itself cause it to slice through the surface, never to return.

It was many years—two centuries, actually—before anyone took the phenomenon as seriously as Spallanzani, and took the time to observe stones as they skipped and describe what they saw. Finally, in the April 1957 issue of *Scientific American* magazine, a retired professor of English, Ernest Hunter Wright, described his own unique efforts at stone skipping. Out on the beach one day, with the water too rough for good skips, Wright tried skipping his stones instead along the damp, hard-packed sand of the beach. When he examined the marks left by the stones he was befuddled: "I think I must have been as astonished as old Crusoe on beholding the first footprint of his man Friday! The first bounce of the pebble was only four inches long; the next was nearly seven feet; then came another short hop of only four inches; then a leap of about five feet; then again the four-inch hop, and so on for seven big hops punctuated by the four-inch ones."

Wright's observations run counter to anything anyone who's ever skipped stones on water has seen—with rare exceptions the distance between each skip lessens until the stone finally runs out of steam and sinks. Yet he was convinced by his time on the beach on that rough-water day that what he had seen must be what happens: "Now I fancy that the same thing happens on the water, though in the water there is no imprint left to tell the story." Well, he *was* an English professor, after all. Wait: that is too harsh. He did have the scientific instinct. Wright proposed that recording the flight and impacts of a thrown stone would be the right way to go, and he even consulted several physicists "of high distinction" for an explanation of the short-long-short pattern he'd seen, without receiving an adequate explanation. Actually, he said they didn't even answer. Wright did receive some suggestions from others. One was that the stone flipped over at every impact (creating the two marks 4 inches apart), then took off again. Another was that the stone was striking first with its rear end, then tilting to hit again with the front, before soaring away. Wright wasn't convinced by either of these, and as far as I know, no one has explained Wright's observations, now nearly fifty years old.

I guess I can't be categorical about that, seeing as how *Scientific American* received more than ten thousand theories from readers of Wright's contribution. Even so, the magazine did not dabble in stone skipping again until 1968, when they published the results of some new, very neat experiments by a student named Kirston Koths. His experiments included the very thing Wright had longed for: photographs. Photographing a stone in flight is no simple task, and Koths was forced to bring the whole event indoors in order to control it well enough to record it. Instead of natural stones—each of which skips in a slightly different way—he manufactured a set of symmetrical sandstone discs and marked them with a single black stripe on one side and a double on the other, so their orientation in any photograph would be immediately clear. To illuminate the stone in flight Koths used a camera that produced a series of flashes, each of which lasted a fraction of a millionth of a second, or, as he put it, "about as long as it takes an automobile moving at sixty miles an hour to travel the thickness of its paint."

It wasn't easy. His first attempts to replicate Wright's skips on sand failed because the light colour of the sand overexposed the film. He had to go back and dye the sand black. Executing the experiment meant Koths had to turn off the lights, turn on the flash, throw the stone (accurately into the tray of sand or water) and open the shutter, all in a couple of moments. But in the end he had some answers.

In the case of the stone skipping on sand, the short hop that Wright noted was indeed the result of the stone hitting, then turning over and hitting again within that short distance of about four inches, and then taking off again. Sometimes that initial rotation would continue: the stone would tumble as it flew through the air, but sometimes it would stabilize itself and skim right side up until it hit the sand again. Still a curious event, but this is all about skipping on water. Koths had problems here: his flash technique didn't work because splashing water obscured the details of impact. So he switched to high-speed film and captured what were the first good images of a stone skipping on water, but more amazingly (although neither he nor anyone else realized it), he confirmed what Lazzaro Spallanzani had seen two centuries earlier.

Koths provided evidence that a typical stone hits the water trailing edge first (because as you-know-who had said, most skipping stones are thrown with a slight tilt, up at the front, down at the back), then slides along the water, tilting even farther back as it goes. As it straightens up on its way to a maximum angle of about 75 degrees (for comparison, the Tower of Pisa leans at about an 80-degree angle), a wave of water builds underneath its leading edge (Spallanzani's inclined plane) and the stone then launches itself out of the water again. It's a one-step event, a long hop, followed by impact, followed by another long hop—no Wrightian short-hop pairs here. But, as Koths admitted, "the entire interaction of the stone and the water is quite complex."

In fact describing the impact of the stone on the water the way I just have is misleading—it is easy to forget that while the stone is tilting and the wave underneath is building up, there is still an enormous amount of forward momentum contained in the stone that will carry it—with any luck—through many more skips. Somehow that energy is able to tear the stone free of its contact with the water and actually send it back up into the air, *against gravity,* until it falls and hits the water once again.

It seems reasonable that some of the difference between the stone's behaviour on sand and water is the result of friction: hitting the gritty surface of sand robs the stone of energy to a much greater degree than "bouncing" off the water, and slows the spin of the stone much more, although the stone still retains enough energy to skip several times. (The spin, like the spin of a top, stabilizes the stone in flight.) But the impact force is significant, likely large enough to actually flip the stone over, sometimes over and over. Out on the water things are different: instead of the impact forcing the stone to turn over, it instead pushes up the wall of water that then turns into the launching ramp for the stone. It should be possible to identify some medium that is somewhere between water and sand in consistency with respect to skipping. Then one could find the threshold where flipping the stone gives way to bunching up the medium. If only my man Spallanzani had known about skips on sand, he would have turned out another learned treatise and we'd all be further ahead.

The final instalments in this story, at least to date, were written in the last fifteen years. In 1988, Richard Crane, writing in *The Physics Teacher*, addressed one of the trickiest aspects of stone skipping: the role of spin. A spinning stone is much more stable both in the air and upon impact, and Crane worked through the equations to satisfy himself that spin is all-important. In particular he was interested in precession, the wobble that a spinning object, like a top or even the Earth, makes when it is spinning (see the chapter "Speeding to a Stop"). That wobble is always in the opposite direction to the spin, and in the case of a stone hitting the water, it should, according to Crane, nullify any tendency to tilt and slice through the surface.

Sorting out the directions in which the various forces will act in a situation like this is tricky, but Crane argued that this is the scenario: the leading edge of a thrown stone starts to dip too low; that makes the stone start to tilt to one side. However, that tilt in turn shifts the force being exerted by the water to one side, and that creates precession—a wobble—that restores the upward attitude of the leading edge. How very neat.

Finally, in late 2002, a French scientist, Lyderic Bocquet, author of the lines I quoted at the beginning of this chapter, crunched the numbers and came up with some theory to back up what people had already seen or assumed.

He pointed out that the maximum lift force on the stone—the thing that is going to get it out of the water once it has hit—occurs when it is only partly immersed, because then the lift force is unopposed. Submerging the stone completely would pit the lift against the weight of water over the stone, and it would, as Spallanzani said, "seek the bottom." The lift force depends on a number of factors, including the speed of the stone and its size and weight. Bocquet calculated that the stone should be spinning several revolutions per second to be stable enough in flight to give the maximum number of skips. And what about that maximum number? Well, a stone thrown properly with a velocity of 12 metres per second (about 40 kilometres per hour) should skip, according to theory, thirty-eight times. And whaddya know? That just happens to be the current world record. A slower stone, such as one thrown 8 metres per second (just over 25 kilometres per hour), should skip only seventeen times. But Bocquet argued that the rate of spin was likely just as important and perhaps more demanding: for that world record throw the equations suggest you'd need a spin rate of 14 revolutions per second. There are, however, so many uncertainties that he cautioned that this number should "not be taken literally."

And what will be next? Bocquet and his colleagues want to build a catapult that will throw a stone in the same, repeatable fashion, time after time, making it possible to subject every stage of the flight to scientific analysis. In that, Bocquet is following in Spallanzani's footsteps. Skipping a stone is great fun, and apparently always has been. Much of that fun comes from the fact that every single throw is different. However, that variability makes it impossible to analyze exactly what is happening. Any scientist would want to understand the step-by-step details of the throw, not to take the mystery out of it (as so many people mistakenly think) but to *enhance* the mystery by identifying the sub-mysteries (and inevitably creating a set of sub-sub-mysteries). The investigation of the skipping stone might never end, but I know that Spallanzani would have loved the idea of a catapult.

CURLING

I don't care what sports elitists say about curling; it doesn't matter to me that many of its participants are less than buff, or that they stand around chatting amicably during a game. The fact is that at the highest level it requires incredible skill. What's even more fascinating is that physicists find the sport of curling a great challenge, because the simple motion of the curling rock as it travels down the ice has, so far, defied explanation.

When a curler releases the rock, he or she gives the handle a twist, either clockwise or counter-clockwise, and as the rock slides down the ice, it begins to "curl," to bend either left or right depending on the turn it was given in the first place. A counter-clockwise turn (as seen from above) causes the rock to move to the left, and vice versa. The amount of sideways movement is substantial, up to a metre or more. This is of course what gives the game its subtlety, and what demands the greatest skill. If thrown properly, a rock can curl around rocks that are already in place on the ice, in effect hiding behind them, immune to attack by all except a rock that travels a similar, perfectly aimed curving path.

The entire game is built around the ability of the rocks to curve as they travel, but why they do that is a mystery. In situations that are superficially similar, the exact opposite occurs. For instance, if you turn an ordinary drinking glass upside down and slide it across a smooth surface, like the polished wood of a bar or a kitchen counter (as they do in one of the officially sanctioned events in the World Beer Games), it will travel more or less in a straight line. But give it a twist and, although it does turn, it moves in the opposite direction to a curling stone. If it's rotating counter-clockwise (again, as seen from above), it bends right, not left. So the curve of a curling stone obviously has some absolutely crucial feature that differentiates it from an upside-down spinning glass. The physicists who have turned their attention to this problem believe that feature is the ice, but when it comes to turning that belief into equations of motion that actually predict what you see in the game, the jury is still out.

To get a sense of what's involved, it's worth looking at the physics of the overturned drinking glass. The glass is more similar to the curling stone than it looks, at least from the physical point of view. Although the curling stone is something like 20 kilos of granite, very little of it actually comes in contact with the surface of the ice. The rock itself is 28 centimetres (a shade less than a foot) in diameter, but the underside is concave, with the result that only a thin band—less than a centimetre in width—rests on the ice. In that sense the curling rock is supported on a rim like the over-turned glass.

Remember, though, that their movement when simultaneously sliding and turning is exactly opposite, and explaining this weird result is a crucial first step to understanding curling. The physics of the drinking glass seems to be pretty well understood. There's no disagreement that friction plays the most important role. Friction is what has to be overcome to start the glass on its way, and friction is what eventually brings it to a halt. If the glass is simply pushed straight ahead, it feels the friction more strongly at the front than at the back, so much more strongly that if you push the glass too hard, it will tip over forwards. With that established, the next question is: how does the friction encountered by a *rotating* glass determine its curved path across the counter? That's the rub. (Ha-ha . . . rub . . . friction.)

To follow the science from here on in you have to imagine yourself suspended over the glass as it spins its way along. Suppose that it is rotating clockwise: the spin alters the straightforward more-at-the-front, less-at-the-back distribution of friction around the rim of the glass. Because the front of the glass is turning to the right, it is experiencing at least part of the friction it feels as coming from that direction. By contrast, the back of the glass is turning left, and so it feels friction coming from that direction. However, the fact that the glass is turning hasn't altered the fact that the front of the glass experiences more friction than the back. That means the force from the right is greater than the force from the left, and as a result the glass must drift to the left. Turning right, bending left: that's the case of the overturned glass on the countertop.

It's not easy to write equations describing these motions that actually apply to the real world. For instance, Ray Penner of Malaspina University-College in Nanaimo, British Columbia (one of the three physicists who have published papers on curling and who live in British Columbia), has launched short pieces of plastic pipe along a bench lightly coated with cornstarch (to leave a visible record of the pipe's movement) to confirm that the pipe (standing in for the drinking glass) does turn in the direction and amount predicted by the physics. It does. Not only that, but roughly speaking, the faster the glass is rotating, the more its path bends. Penner and his colleagues conclude that upside-down glasses are sensible; curling is much more mysterious.

Another one of the B.C. curling boffins, Mark Shegelski of the University of Northern British Columbia, has found that the overturned-glass demonstration startles curlers, who think it's magic, or suspect that somehow the glass has been tampered with. Of course it's also a good demonstration because curlers never seem to be far from a long, smooth polished surface and a plentiful supply of drinking glasses.

If the physics of the drinking glass are clear, the physics of the curling stone are anything but. Obviously there's something different about an object sliding on a sheet of ice, but identifying that "something" has proven very tricky. The attempts go back to 1924. E. L. Harrington of the University of Saskatchewan (Who knows? Maybe if he'd lived longer he would have moved to B.C.) published a report that year in the *Proceedings and Transactions of the Royal Society of Canada* describing his efforts to understand the game, and in doing so he kicked off the noble practice of devising elaborate, Rube Goldbergian apparatuses in order to study the phenomenon under controlled conditions. To his credit, Harrington began by observing what real curling rocks do on a sheet of curling ice. In a sense he converted the entire sheet into an enormous piece of graph paper by suspending a measuring stick over the middle of the ice along its entire length (easily 30 metres!), just high enough to allow a rock to pass beneath. He also marked the ice sheet with scratches a foot apart, again running the entire length. It was a labour-intensive study: four people were needed, one to throw the rocks and then follow them, announcing the completion of each half or full turn, another to record the times of those announcements and two more to locate the exact position of the rock at each of those moments.

Harrington didn't stop there. He created a laboratory version of the game so he could examine the effects of friction in detail. The lab set-up was a kind of role reversal: the ice moved, but the rock remained still. Harrington built an ice-covered turntable, about 75 centimetres across, and positioned the rock on it with pulleys attached to measure the forces the rock experienced when an electric motor rotated the turntable. He even went so far as to "pebble" the ice in the same way a sheet of curling ice is pebbled. (More about pebbling later.) By combining observations of stones thrown on the actual sheet of curling ice and the forces measured on his turntable, Harrington concluded that "the stone has less grip on the ice at higher velocities,"—that is, the influence of friction on the rock is slight just after it has been thrown, but grows as the rock slows down. He used that idea to explain curl. He drew attention to the fact that the two sides of a rotating rock are moving at different speeds *with respect to the ice,* just like the inverted drinking glass. This is sometimes hard to visualize, because obviously the whole rock is moving forward at exactly the same pace—it isn't leaving the slow bits behind. But again, if you imagine hovering over a rock that's turning clockwise, and pick an arbitrary spot on the right-hand edge of that rock, that spot, while moving forward with the rest of the rock, is actually moving backward with the rotation. The net effect is still forward movement, but it's slower than if the rock weren't turning. As soon as the spot gets to the other side, the situation reverses: the spot moves forward with the rock and additionally with the rotation. So it's justifiable to think of the two sides of the rock as moving at different speeds.

As long as the rock is moving forward quickly enough compared with the rotation, those differences are slight. But as the overall speed of the rock slows, the discrepancy between the speed resulting from forward motion and the speed resulting from rotation grows (rotation isn't slowing as much). That means for a clockwise rotating rock, the velocity of the right side is now much less than that of the left.

As Harrington had noted, the friction becomes greater as speed slows. That means the right side of the rock, as it slows, is feeling more friction. However, the left side isn't being held up as much, and that forces the rock to start to turn to the right. Harrington even pointed out that sometimes, if the forward speed of the rock is nearing zero, the difference between the two sides of the rock will be so great that one side will actually stop, and the rock will suddenly pivot, "forming a sort of hook at the end of the path of the stone."

It is so nice to know that this problem was all sorted out three-quarters of a century ago. Or was it? Ray Penner agrees with Harrington's findings, although he's careful to qualify that agreement by adding that for this frictional left/right mechanism to work, there has to be a particular kind of friction involved, called "stick-slip." It's the same kind of friction that operates when a violin bow is dragged across the string—it may sound smooth, but in fact at the micro-level there is an endless series of stops and starts, sticks and slips. With stick-slip friction operating at the front of the curling rock, it will pivot around and move in the expected direction. Penner argues that the curl of stone is the result of a left/right asymmetry of the kind that Harrington identified nearly eighty years ago.

But not everyone agrees that this is the solution. In the late 1980s the above-mentioned Mark Shegelski proposed that curling involves not a right/left discrepancy but a front/back one. If the front of a rock experienced less friction than the back—just the opposite of the drinking glass—then it would curl in the opposite direction to the glass. He pictured the weight of the stone exerting enough pressure on the ice to melt it momentarily. The friction between the bottom of the rock and water is much less than that between rock and ice, and so when the ice is melted the rock slides more easily. But remember, the only part of the rock that is in contact with the ice is a thin ring around the bottom. Shegelski suggested that, depending on the speed of the rock, more or less of that ring would be sliding on water and experiencing less friction. The distribution of liquid water around the ring would be determined by the speed of the rock and its rotation: at slow speeds the rock, as it rotated, would drag water from the back to the front, creating a frictional imbalance that dictated the direction the rock would move. (That Shegelski deserves recognition as a true curling physicist is demonstrated by the fact that, while he hasn't scooted pipes across cornstarch or built a giant turntable, he has built a patented device that in effect alters the diameter of the rim of a standard rock, to see how that influences the amount of curl.)

But these days even Shegelski isn't sure that his original model is the correct one. Ray Penner doesn't buy it—he built the equivalent of the front half of a stone to see if it was easier to drag across the ice than a full rock . . . it wasn't, although if the front half of the rock were experiencing less friction, it should have been.

Others argue that melting the ice can't be the answer. Mark Denny is another physicist (living in B.C.—where else?) who has addressed the curling problem. He contends that the rock simply can't produce the enormous pressures necessary to melt ice, and that even if it did, the pebbling of the ice surface would make that melt-water irrelevant. Here's his argument. A sheet of curling ice is not perfectly smooth. Once flooded, it is layered with a surface of tiny droplets, creating a pebbled surface. The result is that the rock glides on top of an endless series of bumps, just another deception of the sport. A glance at a curling stone gives the impression of a lump of rock solidly anchored on the ice, but in reality the ice is a mosaic of tiny hills, and the only part of the rock touching the tops of those hills is a band a few millimetres wide. That massive rock is only barely in contact with the icy surface underneath.

Denny argues that even if the pressure of the rock melted some of those pebbles, the water released would simply flow into the valleys between, and the rock would continue travelling on unmelted ice. The other curling physicists aren't deterred by this argument, contending that a layer only a few water molecules thick would serve to lubricate the rock/ice surface, and that this water would exist for mere fractions of a second anyway, much too little time to allow any kind of flow. Denny is ready to accept that the rock might warm the ice, short of melting it, and that this alone would lower the friction between the two, but the warming vs. melting debate is, as are most of them in this case, unresolved.

Denny has come up with his own idea of what might be happening. He suspects that, rather than melting the ice, the rock chips away at the protruding pebbles on the ice, and that these turn into debris that gets caught up under the leading edge of the rock. Some of the larger pieces would simply be swept aside (he calls it the "snowplow" model), some tiny ones would be crushed almost immediately, but others of intermediate size would be trapped under the rotating leading edge and carried around to the side. Depending on the size of these bits, how far they're carried around under the rock, and whether or not they survive as ice or melt into water, two opposing frictional forces could be developing, one that would move the rock to the left, the other to the right. Denny is sure that equations point to the correct effect: ice debris moving a counter-clockwise rotating rock to the left.

Although no one has actually witnessed these ice chips, Denny argues that there is circumstantial evidence for them. Anyone who has ducked his head down close to the front of a rock as it's moving up the ice has heard the odd growling sound that it makes. Growling, Denny says, because of the ice that is being ground up as it moves. Also, a light touch on the rock as it's moving reveals that the rock is vibrating as it slides. Denny also points out that every rock that's thrown accumulates debris around the rim of the thin band on the bottom of the rock that is in contact with the ice, debris that could be the result of snowplowing down the ice. (It is an interesting theory, but as far as I know Denny hasn't actually *built* anything to demonstrate it, and that does appear to be the sine qua non of this area of research.)

There's no shortage of theories about why curling rocks do what they do. Few of them, however, come equipped with convincing equations that show them to be correct. The number of articles published on curling in the *Canadian Journal of Physics* in the recent past (ten in half as many years) is a testament to the difficulty of the problem, and there are important features I haven't even mentioned yet. For instance, you might expect, given that turning the handle is what makes the path of the rock bend, that turning it faster might make it bend farther. Wrong. There is *some* relationship between the two, but it's subtle and not yet well characterized. Even though anything over a couple of turns seems not to increase the bend, and adding even more seems to decrease it, Mark Shegelski has had curlers spin the rock so that it turns *eighty* times as it travels down the ice, and such rocks bend a prodigious 2 metres, double the usual distance. Another mystery is the relationship between forward motion and rotation. Does one stop before the other? Can you have a rock that isn't going anywhere but is still turning? And what about vice versa?

And then there's sweeping. Sweeping makes a rock go farther and also reduces the amount of curl. I once heard a theory that, especially in the old days of corn brooms flailing wildly in front of the rocks, the brooms were creating a partial vacuum and sucking the rocks forward. I think the one thing that curling physicists would agree on is that this is not the case. The rock is just too massive, the vacuum—if there is one—too tenuous. But beyond that, who knows? Is the broom beating down the pebbles of the ice? Is it heating the ice, even to the melting point and thus creating a thin layer of water for the rock to slide over? Mark Shegelski once suggested that if sweeping melted some of the ice at the front of the rock, that would reduce the frictional difference between front and back that generates curl, explaining why sweeping reduces the amount a rock will curve.

And so, here in the twenty-first century, we're not really sure why curling rocks curl. Here's what we do know: it has something to do with friction; that friction somehow turns rotation into bending; sweeping is helpful; physicists who investigate curling build cool apparatuses; and apparently you don't have to know the physics to be able to play the game.

Time Passes . . .
Faster

If you're in your teens, this chapter won't make much sense to you, but you had better read it anyway so you'll be prepared for what's in store for you. It's about an experience that everyone has as he or she gets older: as the years go by, time passes faster and faster. You don't really have to be *old* to notice it, but the effect becomes more dramatic as the years go by, so that you eventually find yourself saying, "I can't believe it . . . it's almost the middle of 2004 and I'm just getting used to 2003 . . ."

Almost everyone experiences this phenomenon; the question is, why? One straightforward suggestion that has come up again and again is that our sense of the passage of time is influenced directly by how much time we've experienced in our lives. For instance, one year for my eleven-year-old son, Max, is about 9 percent of his total life; that same year for me is less than 2 percent. If the feeling of a year is based on how much of our total experience that year represents, then obviously, because it is relatively more event-filled for him than it is for me, it will seem to pass by more slowly for him than for me. But suggesting that this explains why time accelerates through one's life is just that: a suggestion. There have to be some data to back it up, and there are.

The single most interesting piece of work on this issue was published by Robert Lemlich of the University of Cincinnati in the mid-1970s (which of course seems just like yesterday to me). Lemlich proposed that one significant adjustment had to be made to this idea that the apparent time of passage of a year is related to the number of years a person had already lived. Lemlich argued that since it's all subjective anyway, the apparent year should be related not to chronological years but to subjective years. In other words, your estimate of the length of the year that has just passed should be compared not with the actual number of years you've lived but with your subjective sense of the length of your life. That each passing year seems shorter means that Lemlich's proposal has some disturbing implications.

Being an engineer, he took a no-nonsense mathematical approach to the question, and came up with a set of equations. For instance, let's assume you are a forty-year-old who will live to the age of eighty. Lemlich calculated that time would be passing by twice as fast now as it did when you were ten, which probably feels about right. Remember how long summer vacation seemed to last, and how the return to school in the fall felt like you were turning a whole new page in your life? But the extension of this calculation might not sit quite as comfortably. The numbers tell you that you're halfway through your life, but because time seems to be passing ever more rapidly, Lemlich's math suggests that you will *feel* something quite different. In fact, he calculates that at your age of forty you have already lived—subjectively—71 percent of your life. It gets worse too: by the time you're sixty, even though you have twenty years remaining, those twenty years will feel like only 13 percent of your life.

These numbers are shocking enough (and you may well disagree with them, but before you do, ask your grandparents how *they* feel), but they take on an even more bizarre twist when you extrapolate them back and ask the question, at what point in our lives do we feel we're at the halfway point of life? The forty-year-old that has been the model of this exercise would have felt as if she were at the halfway point when she was . . . just twenty. I don't mean by this that at twenty she felt half as old as forty; I mean that when she was twenty she had already experienced half her total subjective life.

These numbers are amazing and slightly unbelievable, but Lemlich backed them up with some experiments. He asked a group of students and adults to estimate how much faster time seemed to be passing in the present in comparison to when they were one-quarter or one-half their present age. His theory predicted the answers almost exactly: time seemed to these subjects to be passing twice as fast as when they were one-quarter their present age and 1.4 times faster than when they were half their present age.

Lemlich's theory has been tested in a number of other situations, and although not every attempt at replication has supported his math, many have. There is even one intriguing confirmation from more than two hundred years ago of the part of his thesis that suggests we've experienced half our life while we're still very young. Sidney Ross at Rensselaer Polytechnic Institute came across the following passage written by England's poet laureate Robert Southey in 1837: "Live as long as you may, the first twenty years are the longest half of your life. They appear so while they are passing; they seem to have been when we look back on them; and they take up more room in our memory than all the years that succeed them." Southey continues by referring to an American teacher, Jacob Abbott, who narrowed the age down to between "fifteen and twenty." Ross points out that using seventy as a reasonable lifespan for the time, Lemlich's equation gives seventeen and a half as the point marking half a subjective lifetime, an age that tallies exactly with Abbott's and Southey's estimates.

But *why* would time seem to pass more quickly as we age? It may not be simply that each year represents a smaller percentage of the total. As we get older, our biological clock may run slower and slower, and if it does, then external events would, by extension, appear to be passing by faster and faster. A crude—and exaggerated—example would be that something that formerly seemed to take an hour would, when you were much older, last only half an hour by a slower biological clock.

We do have biological clocks in our brains, and there is some evidence that they can be slowed. In fact we have several clocks, most of which are probably not involved in the apparent acceleration of time. These include the clock that tracks the twenty-nine days or so between menstrual periods and the set of daily, "circadian" clocks in various organs that measure twenty-four-hour periods. Both are somewhat imprecise, and both can be reset (the circadian clock by the day-night cycle, the menstrual clock by many factors, including the presence of other cycling females). However, because these clocks measure time independently of what we're doing or thinking, partly because we aren't aware of their speed and partly because they don't seem to slow appreciably anyway, we can ignore them. They're merely timers that help the body function efficiently.

At the other extreme of the time spectrum is a clock that allows us to estimate the passage of small packets of time, moments that last on the order of seconds. Even though it's not known exactly how this works, many experiments have revealed that it's more complicated than the clock analogy might lead you to believe. After all, if it is just a clock ticking in your brain, who or what is watching it and counting the seconds? There has to be something more.

Imagine you're asked to listen to a musical note, then after it ends there is a pause, then the note is played again, but this time you have to say "Stop!" when you think it has lasted as long as it did the first time. It's a simple enough task, but there is a minimum set of equipment you would need in your brain to be able to do it accurately. One is a timepiece, some sort of biological clock; a second is something that can use that clock to judge the length of the note; and a third is something that can re-create that length while the second note is playing—that's the only way that you can know when to say "Stop." Any clock in the brain that allows you to estimate the passage of time needs these three components.

Experiments like the one I just described have some curious results. Most of us overestimate the duration of very short lengths of time, underestimate anything much longer than that and are accurate only within a very narrow range between the two. In some cases the accurate range is right around six-tenths of a second, in others it is three seconds, depending on the design of the experiment. This pattern of results has been consistent from the mid-1800s to the present. There were suggestions more than a century ago that the two different kinds of errors hinted at the existence of two different timers, and today those hunches have been proven right. There is solid evidence that more than one clock is running in the brain, or at least that more than one part of the brain is being recruited to tell time. Scientists don't agree completely about which parts are involved or what their exact roles are, but in outline the brain's timers work something like this:

Tiny intervals of time—fractions of a second—are tracked by centres buried deep in the brain. These timekeepers operate at levels below our awareness—they are not part of what's called the "thinking" brain. One of them is an area that is affected early in Parkinson's disease; as a result, Parkinson's patients are inaccurate at judging these short time intervals. The neurotransmitter molecule dopamine (the chemical that allows neurons in this part of the brain to "talk" to each other) is crucial to accurate timing: it is depleted in Parkinson's disease and can also be boosted or depleted by a variety of drugs. Marijuana lowers dopamine and slows the clock so that time speeds up. (If you think a slowed brain clock would logically make time seem to pass more slowly, think of it this way: if your slowed clock now estimates a second as being only half a second, then twice as much as usual seems to be happening in that interval, and therefore time will seem to be speeding up—what seems like one minute will actually be two.) Conversely, cocaine raises dopamine levels and speeds the clock.

If you're questioning the usefulness of a clock that is capable of judging intervals of time that are less than a second, question no further. Without such a timer you wouldn't be able to play Jeopardy: rapid and accurate movement of your finger to the buzzer requires extremely accurate timing and coordination on short time scales, exactly what this timer provides.

If you're trying to estimate anything longer than a second or so, you recruit different areas of the brain, areas that are concerned not so much with timing as with paying attention and remembering. In these situations, the cortex, the elaborately folded surface of the brain, is involved. The right parietal cortex, the part of the brain just behind and below your right ear, and the right frontal cortex, an area tucked in behind your right eye, have been identified as playing key roles in judging time intervals of a few seconds. The parietal cortex is known to be important for paying attention: in this instance, attention is important for keeping track of the number of ticks of the clock that accumulate, for instance, while a tone plays. The frontal cortex is necessary for what neuroscientists call "working" memory (better known colloquially as short-term memory), which in an experiment like the one above is crucial to the ability to compare the length of a preceding tone with the one that is being played. Patients with damage to the right underside of their frontal cortex grossly underestimate the passage of a few seconds, but those with damage on the left side tend not to make those errors.

But it's not a clear-cut story yet. The right side of the brain may have been identified as being crucial to timekeeping, but there is a poignant medical case that illustrates just how important the left hemisphere is as well. The patient was a sixty-six-year-old man who was admitted to hospital in Dusseldorf with a tumour on the left side of his frontal cortex. He complained that everything seemed to be happening with what he called "accelerated" motion. He had had to stop his car by the side of the road because traffic seemed to be rushing at him at an incomprehensible speed; it looked to him like a time-lapse movie. The already break-neck speed of television was intensified so much he was unable to continue watching. He felt that life was passing by too quickly for him to cope, and he abandoned his previous activities and hobbies and curtailed his social life.

When asked to estimate 60 seconds, this man's guesses averaged 286 seconds, more than four and a half minutes. That was the amount of time that seemed like a minute to him—no wonder the world seemed chaotic. Imagine four minutes' worth of traffic passing by in what seems like a minute. Yet at the same time he didn't fall further and further behind: he knew the date and was out only about an hour in his estimate of the time of day. But that grossly inaccurate estimation of the length of a minute—think of a VCR running at fast forward—suggested that his internal clock might have been damaged by the tumour and slowed dramatically.

This unfortunate patient suffered from an extreme version of what seems to happen to all of us, gradually, as we get older. It might be that our timer fails on two counts. Remember, a clock not only needs a timekeeper but something additional that knows when to start counting the ticks from the timekeeper and when to stop, and that requires paying attention. If you can't pay attention to the flow of information, you might miss some of the ticks of the clock, and your sense of time will slow as a result. In addition, the timekeeper itself might slow, just as so many other processes in our bodies run down with age.

Fergus Craik and Janine Hay at the Rotman Research Institute in Toronto designed an experiment to reveal just what is happening to our sense of time with age. The experiment had two parts: the first involved viewing a set of bright patches on a computer screen, the second, estimating the passage of time. The patches on the screen had several different properties that could be varied, including shape, colour, size and position on the screen. It was a complex picture, because a single patch could be any one of seven different shapes, sizes, colours, horizontal and vertical positions on the screen. Participants had to identify one or more of those characteristics in a ten-second window. Faced with just one—say, the shape—the task was easy; faced with four or five, it was much more difficult.

They also had to judge the passage of time, in two ways: one occurred when the computer flashed the question "How long have you been working on this trial?" and the other required that the subjects estimate on their own when a certain period of time, such as a minute, had passed. There were two groups of participants: those between ages eighteen and thirty-two, and those between sixty-three and eighty-three. The experiment had a beautifully revealing design: if older people have a slower clock, that would show up in their time estimates; if lessened attention was the problem, that would be clear when they made time judgments during the most challenging parts of the screen test.

The results? Time does seem to pass more slowly for older people. They underestimated the time that had passed when asked "How long have you been working on this trial?" with their guesses averaging around thirty seconds when in fact sixty seconds of real time had elapsed. They also erred in stopping work on the visual task at about 120 seconds, thinking that just a minute had passed. In both cases their timers seemed to be running at about half the speed of the younger group's.

Was it a lack of attention or the physical slowing of a clock? Craik and Hay concluded that the difficulty of the computer screen task had had no significant effect on the older adults' estimations of time, so at least on the basis of this experiment, it seemed that the adults' ability to attend to their clock had not been compromised by occupying their brains with a simultaneous task. The clock was just slower, period.

It remains only to tie the two threads together. The years pass by ever more quickly; our clocks are slowing down. Does the latter completely explain the former? I'm tempted to say yes, but I'm afraid the question is still a little too complicated for that. Neither process is steady and inexorable; both are thrown off track by life events, chemical changes or the combination: life events that trigger chemical changes. All I know is that, clocks or not, I'm determined to enjoy the 19 percent of my subjective life that I have left.

Acknowledgments

Researching a book like this is best done by someone who's meticulous, determined, borderline neurotic and, most important, loves libraries. I've been fortunate when writing books to have been able to find people just like that, and this time around was no exception. Tiffany Boyd did most of the research and always managed to find exactly the right sources, sometimes annotated with her own funny, though acid, comments. I know from playing that role myself for some of the chapters that it could never have been easy. "I'd like to write something about time passing faster as you get older," I'd say, and she'd have to get the goods.

I have quite likely forgotten some people's specific contributions to this book (I'm sorry). These might be people like Penny Park, Adriane Lam and others in the *Daily Planet* office at the Discovery Channel, who get wind of my obsession with some topic and surreptitiously drop articles on my desk. It's a luxury to work in an office (especially in television!) where everyone's talking about science.

Susan Folkins at Penguin Group (Canada) helped turn this manuscript into something entertaining, Shaun Oakey made it readable, and Michael Schellenberg's patience allowed him to wait months for its delivery. My daughter Amelia, her friend Lizzie Barrass and Lizzie's parents are responsible for the crucial experiments on tumbling bagels.

Dan Williman undertook the extraordinary and difficult job of translating Lazzaro Spallanzani's treatise on skipping stones from formal Latin to English—Spallanzani's thoughts made that chapter.

I'm always pleasantly surprised at the amount of time research scientists are willing to take to answer the kinds of questions I ask. Among those who took the time to explain and illuminate their research were Allison Sekuler, Richard Block, Fergus Craik, Oliver O'Reilly, Sid Nagel (a scientist who has the wonderful knack of making crumpling paper and dripping honey profound), Eric Schwitzgebel, Michele Holbrook, Michael McBeath, Bruce Lyon, Karen Wynn, Ray Penner, Mark Shegelski and Mark Denny ("the curling club"), Russell Bernard and Peter Killworth of six degrees fame, Kielan Yarrow, David Greene, Richard Wiseman and Marilyn Schlitz. I thank them all.

Index

Abbott, Jacob, 192–93
acceleration
    judging, 163–64, 166
    of time, age and, 190–98
acetylcholine, 132
aerodynamics, 16
    ideal in, 156
    low-budget, 152, 153
    of maple seeds, 153
    of samaras, 155, 156–58
    seed's wing length and, 154–55
    weight distribution and, 158
age
    chronological *vs.* subjective years
      and, 191–93
    passing of time and, 190–98
Ailanthus, 159
air
    drag, spinning coins and, 101
    flow, falling samaras and, 157–58

Alfred Nickles Bakery, 13
algae, chlorophyll and, 146
Alverez, Gonzalo, 34
ambidexterity, 32
American coot, 43–45
American Journal of Physics, 161
animals. *See also* birds; fish; insects
    counting ability of, 43–47
    eyes of, 127, 128–29
    mathematical ability of, 49,
      50–51
    mental rotation abilities of, 90
    staring and, 122
anthocyanins, 147–51
    insects and, 148, 149
archer fish, 167–69
Archetti, Marco, 148
art, works of
    baby-holding in, 33–34, 35, 41
    human attractiveness and, 132–33